try (1899). "I passed away the time thinking of my own
-place...." Courtesy of The Kendall Whaling Museum,

LIFE ON THE WAVE

5/92 Astoria, Oregon Columbia River Maritime Museum

Charles L. Newhall, nautically attired, years after his service on whaling and merchant ships. Courtesy of the Kendall Whaling Museum, Sharon, Massachusetts.

THE ADVENTURES OF JACK: OR, A LIFE ON THE WAVE

BY

CHAS. L. NEWHALL

" 'Tis strange, but true, for truth is always strange,
Stranger than fiction, If it could be told,
How much would novels gain by the exchange!
How differently the world would men behold!"

BYRON'S *Don Juan*.

Edited, with notes, and introduction and an afterword

by KENNETH R. MARTIN

Ye Galleon Press
Fairfield, Washington
1981

Newhall, Charles Lyman, 1834-
The adventures of Jack, or, A life on the wave.

Running title: Life on the wave.
Reprint. Originally published: Southbridge, Mass. : The author, 1859.
Bibliography: p.
Includes index.
1. Whaleman—Massachusetts—Biography. 2. Newhall, Charles Lyman, 1934- . I. Martin, Kenneth R., 1938 . II. Title. III. Title: Life on the wave.

SH381.N49 1981 639'.28'0924 [B] 81-19829
ISBN 0-87770-263-2 AACR2

Errata

In the index those items listed as page 88 should be numbered page 7.

TABLE OF CONTENTS

EDITOR'S INTRODUCTION

Charles L. Newhall's *The Adventures of Jack* – is one of many non-fiction whaling narratives published in ante-bellum America. Such works are the product of an era in which the expanding, labor-intensive whaling industry recruited large numbers of young landsmen with somewhat more than rudimentary educations.[2] Some of these were gentlemen to whom a whaling voyage seemed to promise romance and adventure.[3] Others, like Charles L. Newhall, were intelligent, literate youngsters forced by circumstance into harsh trade.

The expansion of the whaling industry was simultaneous with the Age of Democracy in the United States, an era in which the arts were felt to be within reach of intelligent amateurs.[4] Accordingly, talented (and not so talented) amateurs, convinced that their stories had wide interest, published autobiographical narratives. Most of these sank like stones, the victims of their literary pretensions, artless styles, insufficient blood and thunder, or the American reading public's growing literary sophistication.

Time has been kind to a few of these volumes, Newhall's included. However commonplace Newhall's adventures may have been in the 1850's, they now have novelty and interest. Today, Newhall's versifications seem more endearing than stilted; and when one takes into account the author's youth (he was twenty-five when *The Adventures of Jack* appeared), the book seems a real accomplishment.

Newhall may have kept shipboard diaries from which he later drafted his book, for with a few exceptions the incidents and details he relates are quite accurate, more so than usual for someone writing a memoir. In any case, *The Adventures of Jack* has come into its own as an entertaining, authentic American whaling account.

This new edition of Newhall is reprinted exactly as he wrote it, except that ships' names are italicized and a few confusing typographical errors have been corrected. Newhall's quaint spellings of geographical locations and his use of obsolete geographical place names have been retained. Examples thought to be potentially confusing to readers are clarified in the notes, along with a few esoteric nautical matters.

BIOGRAPHICAL NOTES

Charles L. Newhall completed his autobiography at the age of twenty-five. When his book appeared in 1859, he was convinced he had swallowed the anchor. He was established as a job printer in Southbridge, Massachusetts, and *The Adventures of Jack* was proof of his professional change of heart.[37]

It was not to be. When the American Civil War broke out, Newhall enlisted in the United States Navy, serving on board the U.S.S. *Minnesota* in the eventful blockade of Hampton Roads. His whaling days had accustomed him to life in a maritime trade overdue for modernization, but the navy was another matter. Newhall's ship sustained an attack by the new Confederate iron clad *Merrimac* and another by torpedo.[38] Newhall's printing experience served him in good stead. He was assigned to a shipboard press, where he turned out "letter heads, envelopes and all general orders to the fleet," and where his advantageous location on board allowed him to lengthen his treasured list of celebrated acquaintances by such names as "President Lincoln, Vice-President Hamlin, Secretary of State Seward," and many others.[39]

Newhall was discharged in 1867 after two hitches in the navy. Meanwhile he had married Josephine Robinson, daughter of a whaleman. The Newhalls had three children before 1872, the first of whom died at birth and the second at age two.[40] Newhall worked at the Government Printing Office in Washington, D.C., until 1873 and then pursued several unsuccessful private printing ventures. After 1878 he was employed as "a spare hand" for Southbridge's three newspapers and was active in freemasonry, "the Grand Army of the Republic, Kearsarge Association of Naval Veterans, Boston Council of the American Protective Association of the Universalist church, and the Society of the Sons of the American Revolution."[41]

In 1889, at age sixty-five, Newhall published *The Record of My Ancestry*, a geneology from which the above biographical facts are taken. Summing up an eventful but undistinguished life, he saw himself as a man of "very checkered experience."[42] He had mellowed and grown modest over the years, but he took obvious pleasure in recalling and recounting the nautical adventures of his youth.

AUTHOR'S INTRODUCTION

I would, with pleasure, introduce this work to the Public, as being a true history of my travels and adventures, as I have traveled over most of the civilized world, and have seen and endured many hardships. Many, no doubt, have made a resolve, as I did myself, in their youthful days, to lead

"A life on the Ocean wave;"

but to my youthful readers I would say, that you know not what a sailor has to endure, if you did, you would remain at home; you know not the tender mother's love, if you did, you would regard her grief. And, fathers and mothers, brothers and sisters, if you would keep your loved ones by your side, you should endeavor to cherish them with the most tender care. This may point out to you the dangers and trials that are experienced upon the briny deep. And I would say— —

Roll on, thou dark and deep blue ocean—roll!
Ten thousand fleets sweep over thee in vain;
Man marks the earth with ruin—his control
Stops with the shore; upon the watery plain
The wrecks are all thy deeds, nor doth remain
A shadow of man's ravage, save his own,
When, for a moment, like a drop of rain,
He sinks into thy depths with bubbling groan,
Without a grave, unknell'd, uncoffin'd, and unknown!

C. L. N.

THE

ADVENTURES OF JACK;

or, a Life on the Wave.

CHAPTER ONE

"I've wander'd on thro' many a clime where flowers of beauty grew,
Where all was blissful to the heart and lovely to the view—
I've seen them in their twilight pride, and in their dress of morn,
But none appear'd so sweet to me as the spot where I was born."

I WAS BORN IN THE YEAR 1834, IN THE TOWN OF Spencer, a small town, situated in Worcester Co. fifty miles from Boston; and strange to relate, that my father, grandfather, and great-grandfather were born in Spencer. At an early age (five years old) I had a determination to see the various countries, and from that time until I finally succeeded, I was continually studying my old geography; and with a little ill-treatment on the part of my unfaithful paternal protector, tended to strengthen that determination. When at the age of fourteen years I procured a writing from him to certify that he had nothing more to do with me. I remained in town nearly a year and then went to Boston, and was fortunate in getting employment as clerk in a large boot and shoe store opposite Oak Hall. Here I remained until I got a chance for going on board of a ship, which was in about three weeks' time, when I embarked on board of a whale ship (as many of our young men do) bound to the Indian Ocean.

The next day I went to Fairhaven, Mass., the ship being there fitting for sea. And just before the ship was ready, I wrote a letter to my parents informing them that I was about to sail from my native country, and should probably be absent three or four years. At the same time I knew how they would feel when they should receive the letter, and those same thoughts haunted me for many a long day after we sailed. At length the ship was ready to sail—the old ship *Leonidas*,—the crew all on board, and the ship sailed October 13th, 1849. I was cabin boy. As we sailed down Buzzard's Bay, I gazed upon the rocks, and with a sorrowful heart I bade my native land farewell. And then—

The sails were fill'd, and fair the light winds blew,
As glad to bear me from my native home;
And fast the white rocks faded from my view.
And soon were lost in circumambient foam.

I was sea-sick nearly three months, and often times thought of home; but this was only a commencement of my troubles.

The Captain was a Cape Codman, and a hard case too.[5] Before we went on board he told us the fare would be "beef and bread one day, and bread and beef the next for a change." And we found his words to be true. However, time passed on slowly, as the crew were rigging over the old ship which was nearly fifty years old.[6] The crew labored very hard, and the extremely bad provisions made it still worse. We did not see a ship until we had been out about one month, which was another of our tribe, when near the Western Islands.[7]

When about three months from home, the man at the masthead, one pleasant morning, cried out "Land ho!" and as we came near it, we found it was merely a rock—a small island called Trinidad, in the South Atlantic.[8] Here we lay "off and on" a day or two fishing for codfish, but we did not succeed, so we passed on. At that time I had recovered somewhat from sea-sickness, and was able to go on deck; but still I thought of home; I thought of the many sleepless nights of my maternal watcher; I thought of unwary boys who might follow me in my path of recklessness.

Some of the crew, of which the largest portion had never been on board of a ship before, considered it a very hard task to go up to the masthead to look for whales, but if they said a word they would soon have a piece of a rope to help them along.

Sundays and Thursdays we had the sailors' favorite dinner—a sort of a pudding, which they call duff, made of flour mixed with salt water, and boiled in a bag in a kettle of salt water.

And onward we sailed looking for monstrous big fish, when we came to the island of Tristan de Achuna, a few degrees south of the Cape of

Good Hope, which is inhabited by a few civilized beings, of whom our captain purchased swine and potatoes.[9] We lay "off and on" there for a few days, but we could not get on shore.

Not many days after we left the last mentioned island, we had the pleasure of capturing a whale, of which I will give a description: In the first place we lowered three boats, which was all we had, our ship being a small one—some carry four and five, and sometimes six boats—and pulled for the sea-monster. After two hours hard rowing in the boats we succeeded in driving two harpoons into him about two feet before he began to feel it much, and then he started off at full speed, first going down about two hundred fathoms and then rising again. At last he formed a perfect circle around the ship; and he went round three times, roaring like a lion, and spouting water, reddened with his blood, to the height of twenty feet into the air, sometimes drenching us with it.[10] We expected every moment to be smashed to atoms by his tremendous tail. When he came up to blow we would pull up to him and give him four or five feet of iron in his side, which would tickle him up good; and at last the lance was driven into the right spot, which caused him to turn "fin out."

The next operation was to tow the lifeless monster to the ship. All three boats being fastened together, the hindmost one being fastened to the whale and the others in single file—all pulling together. As we came up to the ship, the sails were partly thrown aback to keep the ship steady until we could get him on board. This being done, we then made him fast to the ship, which was done by means of a large chain at each end.

The next was to take his coat off. A chain was fastened around one fin, and having two large tackles rigged at the mainmast-head, we raised up the monster's fin inch by inch, and at the same time cutting off one lip and hoisting it up, and so turned the whale partly over. When he was sufficiently turned over, we hooked on to the other lip, and then the other fin, at the same time cutting off the head. Then, as the whale turned, the blubber or fat part was cut from the carcass in pieces varying from four to six feet in breadth, which are commonly called "blanket-pieces," and are from six inches to two feet in thickness.

The pieces are hoisted up as high as they can be with one tackle, which is from fifteen to twenty feet, and then the other one is hooked on.

It occupied one day and one night to get our great prize on board; and the next thing before we slept, was to clear away the try-works and to get them to steaming; all the spades and knives were sharpened, we then commenced cutting the blubber, and by means of a mincing machine, at which I had a station, we minced it fine and tried it out. We had two large try-pots which were set in bricks on deck, and would hold one hundred and eighty gallons each. Some larger vessels have three pots which hold from two hundred to two hundred and twenty-five gallons each. As we tried it out, it was bailed from the pots into a large copper vat called a cooler, which holds from three hundred to four hundred gallons. As it cooled it was bailed from the cooler into casks, which hold from sixty to three hundred gallons each.

While a portion of the crew were trying out the oil some were engaged in cleaning the bone and putting it below,—bone that ladies' hoops are manufactured from.[11] After the bone was all cleaned and stowed below, the oil was also put below into other casks; and then the deck was scrubbed with ashes. And at the same time a good lookout was kept at the masthead for more whales.

But I will now continue the voyage. After taking the whale, we proceeded farther on to the eastward, until we arrived near the island of St. Paul, a place which is noted for its fish, as there is a superabundance of them, and they are very fat. I heard an old sailor say that some of them were so fat that others lived upon thc wake of them. We caught about six bushels in two hours.[12]

After leaving St. Paul's, we steered on to the northward; and two or three weeks afterwards we captured three small spermaceti whales, from which we got some teeth. The whales made about eighty barrels of oil, making in all, about one hundred and eighty barrels. We caught occasionally a porpoise, and various other fish, among which was a shark which was caught one pleasant day, as we were fishing for bottom fish. He was hauled upon deck and killed. He measured five feet in length, ten inches cross the head, and had four rows of teeth.

After being from home about eight and one-half months we reached a port—an island called St. Mary's Island in a large bay on the eastern coast of Madagascar. It happened to be in the unhealthy season, and so eager were we to reach the shore, unmindful of the danger that would befall us, so that nearly all the crew were attacked with the African fever. In a few days we were all, with the exception of three or four who did not remain ashore over night, unable to help ourselves, among whom was our gallant captain. He thought himself that had it not been for one of the sailors who he got to bleed him, that he would have been numbered with those below, for the man nearly took his life as he bled him.

That was a hard time for me as I lay in my berth almost unable to move, and could obtain nothing to satisfy my craving appetite, except a little rice and water, but on the poor man's system, that is, "two waters to one rice," and only seven table-spoonsful each day. I tell you what it is, reader, I thought that was hard living.

Notwithstanding all my trials, I passed away the time usually, in thinking of my own dear native land; of the farm-house; of the old-fashioned fire-place, with my parents before it thinking of their lost son, and my little sister by their side mourning the loss of her brother; and that old knotted apple tree by the well, which I very well remembered, as I often sat beneath its shade on a hot summer's day. All of these things I would recall to mind ever and anon as I lay helpless in my berth. And Oh!—

"Thou spot of earth, where from my bosom
The first sweet tones of nature rose,
Where that I cropp'd the stainless blossom
Of pleasure, then unmix'd with woes;
Where, with my new-born powers delighted,
I tripp'd beneath a mother's hand—
In thru the quenchless flame was lighted,
That sparkles for my native land."

After about two months of sickness, and four had died on board of the ship, we persuaded on Capt. Gifford to go into port with us, and accordingly we went to the Isle of Bourbon,[13] a French port, that being the nearest, and four of us were taken to the hospital. We had very good provisions in the hospital—consisting mostly of chickens and good warm bread—which I relished very well; but the greatest difficulty was, we could get it only twice a day—at ten in the forenoon and at four in the afternoon—which, I suppose, made it taste much better. One man died, named Griswold of ________, near Hartford, Conn. He was partially insane, and was continually talking of his good and most-affectionate wife who he had left at home. And Oh! tell me, reader, where is the husband who does not think of his wife when far away; or the wife who does not think of her husband; or a mother who does not think of her son; or son who does not think of his mother? I would answer, I think, with the greater portion of the enlightened community, that there are none who do not think of each other when far away.

Mr. Griswold had blisters all over his body, and sometimes he would pull them off and throw them on the floor. And, at last, his awful misery was at an end—he died; his last words were, "Tell my dear wife that I am no more." His remains were respectably buried. I also had two blisters drawn, of which I shall carry the scars so long as I live. After lying in the hospital about six weeks with tender care, I was taken on board of the ship again, although very sorry to go, but I was still helpless, and was obliged to go.

It was not long after I went on board when the Monsoons came on, which probably the reader is aware that it is a most violent storm, and I began to think that my days would soon be at an end; but alas! and wonderful to say, that the old ship weathered it, and came out safe and sound.

We cruised around near the Isle of Bourbon for a few days, but without success. And then we sailed around to the southward of Madagascar, and up the Mozambique Channel, to a small place on the coast of Madagascar, for wood; but owing to the enormous price of wood at that place, our captain thought best to go a little farther north, in the mouth of a river, where there were no inhabitants, so that the crew might go on shore and cut the wood and bring it on board without leave or license of any one. When we arrived at our place of destination, Capt. Gifford sent a boat's crew on shore for wood, and as there was no more well men than was necessary to go into the boat, the sick list were ordered upon deck to prepare a place for the wood; although we were scarcely able to stand alone. And as we had a place prepared the boat came off loaded down to the water's edge; and as they put it on deck from the boat, it was our duty to throw it into the hold.

One Thursday (duff day) the boat's crew was on shore for wood, and at noon the dinner was taken forward to the forecastle as usual, but none of the well men being on board, the sick ones—Samuel Prescott, James Linard, and two others who dived forward,—thought they had a perfect right to it, and of course took possession of the duff. James Linard devoured nearly all of it. And after that he was kept chained to a post, and fed upon hard bread and cold water,—a very scanty supply. I lived in the cabin, and considered I had a very hard time of it, but it was easy when compared with his.

We finally got our supply of wood, and on the day previous to our departure (Sunday) the crew were allowed to go on shore with their guns, and were gone all day. They caught considerable game, consisting of monkeys, eagles, parrots, &c. And before they came on board they set fire to the woods, which destroyed several acres of wood and timber. There were plenty of wild hogs on shore, but they were very shy. We also saw several crocodiles in the river, but dare not attempt to catch them.

We proceeded then across the Channel to Zanzibar; but we only remained two days, and being at some distance from the town, I had not a very good view of the place.

We then sailed up the Channel, to an island called Johanna, in the northern extremity of the Channel.[14] The inhabitants are descendants of Arabia, and of the Mohammedan religion. The crew all went on shore there. That was the first time I had been on shore since my last sickness, and the third time since I left home. And at that time I was not able to travel much, but I did not travel far before I came to a sugar mill, where they were making sugar. They were very kind to give me the privilege of eating some of it; so that for once in my life I had all the sugar that I could eat. The captain purchased of the natives fifty bushels of sweet potatoes and sixty barrels of yams; also, ten barrels of cocoanuts, three barrels of oranges, and twenty barrels of bananas and plantains.

The cocoa tree grows in warm climate, and is of the order of Palmie. It rises to the height of 80 feet. The bark is smooth, of a pale brown color, and the tree often leans to one side. The leaves or branches are 14 or 15 feet long, about 28 in number, winged, of a yellow color, narrow, straight and tapering. The natives in warm climate often build their huts of cocoa leaves. The nut grows in clusters of about a dozen each, on the top of the tree.

The banana and plantain are nearly the same species, both of which grow in tropical countries. The tree rises from 15 to 20 feet high, with a soft stalk, marked with dark purple stripes and spots, with leaves from six to eight feet long, and from one foot to one foot and a half in breadth. They make a splendid shade, and the leaves are cut in pieces and used as dishes, spoons and cups; also, as parasols for the fair sex. The flowers grow in bunches, covered with a sheath of a fine purple color. The fruit is five or six inches long, and an inch or more in diameter; the pulp is soft and of a luscious taste. When ripe, it is eaten raw, boiled, baked, or fried in slices. They grow in clusters on a stem, similar to grapes, and from ten to twenty pounds to a bunch.

We passed close by the island of Mahe, and others of the Seyschelle Islands, but did not stop. And to Bird Island, a small, low island where the sea-birds go to lay their eggs, and rear their young. A portion of the crew went on shore and brought off about half a bushel of eggs of all kinds and sizes. I had a good feast that time.

We then proceeded to Mauritius, an island at which the vessels going to and from the East Indies often call in for wood and water and provision. I had not wholly recovered from my attack of fever; but notwithstanding, I concluded that I had been long enough on board of the *Leonidas* with Capt. Gifford, and the day the ship was to sail I went on shore with the captain, and as soon as I got a chance I gave him the slip.

I should have stated, that two nights before, one man swam ashore, and the next night—the night before I went ashore,—another man swam ashore. The U.S. Sloop of War, *Plymouth*, was lying there at that time, on her homeward passage from China, so I thought it was a good opportunity for me to go home. However, I remained on shore three days, and each day Com. Vorhees sent a midshipman for me; but as I did not think they wanted me I did not go. And at last a lieutenant came to me and told me that he wanted me to go on board and go home; I went.

While I was on shore I saw three others that went on shore in the boat as I did. We were all taken on board and clothed, so that we might go home comfortable. Two days after I went on board we sailed for the United States (Nov. 1850). On our passage we called at Cape of Good Hope for wood and water, and partially satisfied our appetites with delicious apples, pears, etc., which I had not seen since I left home. On the 18th of December we sailed from the Cape, expecting soon to see "The girl I left behind me."

Then as I sat at the cabin table beside the old Commodore, I thought of the following stanza:

"I cannot think of sorrow now; and doubt
If e'er I felt it—it is so dazzled from
My memory, by this oblivious transport."

Time passed on slowly but easily for me, and on the 28th day of January, 1851, a bitter cold day, we arrived safe at Norfolk, "old Virginia." After remaining at Norfolk one month, when the men from the ship had received their pay, one of them took me on board the schooner *Bounty* and paid for my passage to New York. I arrived there on Washington's birth-day, and there was a roar of cannon heard during the day. I found some fifty men from the *Plymouth*, and they contributed sufficient funds to carry me safe home, where I arrived the next day. And the people flocked around me, and with their thousand and one questions, they soon wearied me, so that I almost wished myself at sea again.

As I was passing along the street from the depot to my father's house, I met a man with a load of hay. He stopped his team and asked me if I had not been to sea.

"Yes, sir," said I; "I have been whaling."

"Well," said he, "I had a boy that went out on a whaling voyage, and have never heard from him since."

That reminded me that I had not written to *my* parents since I sailed, and felt as though I should like to be at home; but he did not detain me long. He asked if I had no curiosities.

"No, sir," said I; "it was all I could do to get home myself without the curiosities."

CHAPTER 2

IN THE PRECEDING CHAPTER I STATED THAT A young man named Samuel Prescott was among the sick. He was very sick at the time I left the ship; and he told me if I should ever reach the United States, to go to Boston and inform his friends about him, for he imagined he never should get there himself; and I afterwards ascertained that he died. I went to Boston as he told me, and spent much time in looking for his friends, but I could find no trace of them.

The thought never entered my mind at the age of twelve or fourteen years that I should ever be in the presence of Emperors, Kings, and Princes; but the time has been when I stood before them and conversed with them, and considered myself as prominent as they were, — as will be seen further on in my travels. They are always in danger, — as Pope says:

"Uneasy lies the head that wears a crown."

From time to time, as I sat by the old-fashioned fireside at home, I thought of the many toils and dangers I had faced upon old Neptune's possession. However, I succeeded in stopping ashore one year, as I was employed in a printing office in Worcester County; but I was wearied of that for the time, and then —

"Like an eagle caged I pine,
On this dull, unchanging shore;
Oh! give me the flashing brine,
The spray, and the tempest's roar!"

So off I went to Boston, and embarked on board the schooner *Eglantine*, as "chief cook and bottle washer." This time I did not write to my parents before we sailed. All was ready, and we sailed on the 10th day of March, 1852, a fine, lovely morning, and ere the shades of eve came over us, we could see nothing save the clouded sky above, and the deep blue beneath us.

After fifteen days sailing we arrived at Port au Prince, Hayti. Sometimes it was stormy and wet, and then I was called to "lend a hand."

When we arrived at Port au Prince, it was in the evening, (Saturday) the town was illuminated, and it seemed to be a very good looking town. But the next morning, we found we had been deceived, and the bright rays of the sun had exposed it, so that our view was altogether different. The male portion of the inhabitants were nearly all soldiers, and clothed in rags; and buildings were mostly low, dirty looking, built of wood, and not painted. There they were getting their breakfast, and cooking it in the street, with about as much wood as I could hold in my hand, for they did not have stoves in that country. One day we were invited to dine with the Emperor, as we were fortunate enough to be there when he was made Emperor, and I thought if that was a specimen of Sovereigns, I had enough of them.[15] We remained there about two weeks discharging the cargo which consisted of salt beef, pork, soap, candles, &c. and taking in a cargo of logwood and coffee.

I went ashore frequently, as I had nothing to do but to cook, and wandered off over the mountains and through the fine forests of logwood, all as beautiful and pleasant as nature could make them; and the orchards of ripening oranges; and by the splendid coffee plantations. One day I entered a printing office, and seeing a vacant composing-stick and some copy which one of the compositors had left a few moments, I went to work setting type in the French language, and I soon found that I could do it very well, and the proprietor of the establishment saw me, and offered me very good inducements to remain there and work for him; but I was so situated at that time that I could not stop there. He was a mulatto, but was an American by birth, and learned his trade in the United States.

At length we sailed again for Boston, and in sixteen days "hard scratching," through thunder storms, hail and rain, we arrived in Boston harbor, although I thought sometimes that we never should see Boston again, for—

"The storm howl'd madly o'er the sea,
The clouds their thunder anthems sang,
And billows, rolling fearfully,
In concert with the wirlwind rang."

It was the 1st day of May, 1852, when we arrived in Boston; that evening Kossuth made an address in that city. On the following evening I slept in that well-remembered bedroom at home.

I could hardly convince my parents that I had been to the West Indies so soon. But there was one evidence which convinced my maternal protector that I had been there, namely, a pair of shoes which I had purchased for her in Port au Prince.

I remained at home during a part of the summer, and the remaining part until about the 1st of October I was visiting my friends and relatives, when the sea-fever came on again as bad as ever. I then informed my parents that I intended to go to sea immediately. My father did not say much about it; but all that my mother could do or say made no impression upon me—all was naught. She wept bitterly, but I heeded it not—I was determined to go. And—

> *"Far I go where fate may lead me,*
> *Far across the troubled deep;*
> *Where no stranger's ear shall heed me,*
> *Where not tear for me shall weep;"*
> *And, like some low and mournful spell,*
> *I whisper'd but one word—farewell.*

Then we parted. My father did not see me go. I went to New Bedford, and in about two weeks I succeeded in getting a chance as steward on board of the ship *Copia*, captain C.M. Newell,[16] for a whaling voyage to the Pacific and Arctic oceans. I wrote to my mother before we sailed, informing her that I was about to leave the United States, and perhaps never to return; also, stating the name of the ship and how long I expected to be away. I well imagined her feelings as she would read those parting lines, and that she would weep when she thought what my fate might be. The *Copia* was expected to be absent about two years and a half, but as I did not ship for the voyage, it was uncertain when I would return. While in New Bedford I boarded with Mrs. Kelly, a widow.

While I was in New Bedford Capt. Gifford arrived from England. He appeared to be very happy to see me, and we sat in a store where we conversed two hours. I told him how I came home, and that the others came home with me. He said he did not expect I should live to reach home again. He informed me that about thirteen months after I left the ship, she was condemned and sold at auction, and that he took the oil on board of another ship, in which he went to England. They did not take a whale after I left the ship. He also told me that Samuel Prescott and James

Linard had died, — Linard was a native of New Jersey, who died about four months after I took my leave. When Linard died, Prescott said it certainly must be his turn next, and so it proved to be, — he died one month afterwards. What a melancholy affair. I afterwards saw the third mate and one of the boatsteerers. They told me that they did not receive a cent for their voyage. Take warning my young friends.

I went on board a day or two before the ship was to sail for the purpose of getting everything to order in the cabin, and prepare myself to perform my duty as a steward. At eight o'clock in the morning on the 18th of October, 1852, we sailed from New Bedford. It was a very pleasant morning, and the sun shone beautifully over the hills. And —

Soft as a bride, the rosy dawn
From dewy sleep did rise,
And, bathed with blushes, had withdrawn
The mantle from her eyes;
And, with her orbs dissolved in dew,
Bent like an angel gently through
The blue pavilion'd skies.

The flags were all hoisted, the sails loosed and set, and in a few short hours we were sailing upon the bosom of the broad Atlantic. I was not seasick that time, and it was somewhat pleasing to me to observe the others who had never been at sea before, more so, probably, because I had been in the same situation myself. But I had a pretty hard time of it, as the cook was a native of the island of St. George,[17] and could speak scarcely a word of English, and knew nothing of cooking; the cabin boy was sea-sick for a week or two, so that I had a double portion of labor. However, I happened perchance to embark with a good easy captain, and occasionally he would assist in making bread, cakes, and pies, as he was an excellent cook. There were thirty-five persons on board, and it was considerable of a task to prepare food for them. On the first evening after we sailed I remained upon deck until nine o'clock watching the full moon gliding along through the cloudless sky; and how beautifully the sweet silver light gilded the waters, as our gallant whip was ploughing through the sparkling waves; occasionally a spray flying over the ship to sprinkle us a little, but I did not care for that. At nine there was not a vessel, or even a trace of land to be seen — naught save the clear boundless sky and the deep blue ocean was in view. In the evening the breeze freshened somewhat, and the main royal was furled. I went below to my berth at nine and slept very soundly until the sun had almost caught me sleeping. I hastened up and got breakfast ready, however, as late as it was, before the captain arose.

I went upon deck, but no land could be seen, — America was lost beyond the horizon. The sun had appeared above the blue waters in all its splendor and glory. Not even a ship was seen until we had been from home one month, when the whale ship *Benjamin Rush* came in sight, and we finally came within hailing distance and conversed about the same time that we did. One duff day as the dinner was carried forward to the forecastle, two gormandizers were determined to devour the entire lot. As it is boiled in saltwater about one inch of each end was liable to become water-soaked, and in consequence of which some of the new-made sailors did not like those end pieces. So as it appeared in the forecastle one of the greedy fellows said to a landsman:

"I believe you don't like end pieces, do you?"

"No," replied the landsman.

And went around to all in the same way, except his friend who was to share it with him, and then said to his friend:

"If no one else will take the end pieces, then of course we must take 'em."

And one of them took out his knife and cut the duff as nearly through the center as he could, thus dividing it into two equal parts; and he then took one half himself, and his friend took the remaining half. Both having end pieces, and still had the whole. The landsmen sat all around on their chests perfectly astonished, but said not a word. They, of course, were obliged to make a dinner of salt junk and hard bread.

The first we saw that resembled land was the Rock of St. Peter, which is situated in about the center of the Atlantic, or somewhat nearer the South American coast, fifty miles north Latitude. It was not more than twenty feet high and about a quarter of a mile in circumference. The cabin boy drew a very good picture of it.

On the following Sunday after we passed this rock, we caught a sunfish, a very large, clumsy looking fish, which we soon dissected. The skin is as rough as course sand-paper, similar, in that respect, to shark skin. We took out the liver, and it filled two common sized pails, which made near a pailfull and a half of clear oil, an excellent remedy for Rheumatism. A portion of the meat is similar to India rubber. We cut out some of it, and it would bound like an India rubber ball. And a portion of it is very good eating, a piece of which I cut off and cooked, and it was good.

We frequently caught a porpoise, and sometimes we would have five or six hung up in a row upon the main stay; and occasionally a dolphin.

As we came down by the Rio de la Plate, it rained continually for a week, until we had passed it; thunder and lightning in no very small quantities.

We passed through the straits of La Maire, which divide Staten Land and Terre del Fuego. The latter, in English, is the "Land of Fire;" but, as we passed by it and viewed its snow-covered mountains in midsummer, I certainly thought it was not very rightly named.

On the 1st day of January, 1853, we were in sight of Cape Horn, and the captain took a bottle which he corked and sealed perfectly tight, then fastened it to a lead with a line attached, threw it overboard and let it remain a few moments, and then it was hauled on board, being filled with pure fresh water. I thought at the time that it was hardly creditable, but it cannot be that my eyes and taste deceived me.

Immediately after passing Cape Horn which is a small island called Diego,[18] it commenced very suddenly to blow a gale from the north-west, the precise direction which we desired to go, and it continued to blow three weeks, which carried us away to the southward as far as 60° South Latitude, and there we were beating and scudding about under short sail. Nearly every particle of clothing of all hands were as wet as they could be, and although so wet, I can say that I did not take cold; and if my recollection serves me, I never caught cold at sea in my life, but have very frequently caught cold when in port, and the next day after leaving port I was perfectly clear of it.

At length, however, the storm ceased, and the wind changed to the southward. We were then as happy as clams at high water. Although as I thought the passage a very long one, it was short when compared with some, for I once knew of a ship that was three months off the cape before she could go round.

As the old *Copia* was gliding along with a fair wind, I thought, and have many times thought, that there never was anything appeared quite as well as a ship with all sail set. In a few days we reached the celebrated island of Juan Fernandez, where Alexander Selkirk was supposed to have been a resident many years.[19] The island is noted for peaches, both for their large quantities and for their enormous size. And around near the coast fish of every description can be found in superabundance. After satisfying ourselves with delicious fruit, and with two boats full of all kinds of fish which we caught in about two hours, among which were seventy lobsters, we proceeded on our voyage.

The next land we saw was St. Felix's Island, which is some distance to the northward of Juan Fernandez. We passed close by it, but it being just at night we did not stop, nor could we have a very good view of the island, although I should judge it to be merely a barren rock.

Not many days after passing those islands, we succeeded in capturing a small spermaceti whale, out of which we produced twenty barrels of oil. And soon after we caught six black-fish, one of which measured twenty-five feet in length, and, producing in all six barrels of oil, which is nearly as good as spermaceti.

One very calm, pleasant day, when in about 5° South Latitude in the Pacific Ocean, we espied, at some distance off, what was supposed to be a very large fish, and as we came near it we ascertained that it was a bone shark, which Capt. Newell supposed to be not less than forty feet in length. He seemed perfectly quiet and harmless, and unapprehensive of danger, darting about with his monstrous tail (I don't know but I should say *narrative*, as the old lady said) partly out of water.

On one fine Sunday morning, while one of the crew was sitting upon the taff-rail fishing, a very large bird came down and took up the hook, which was scitting along in the wake of the ship, and had merely a white rag fastened to it for a bate. The bird arose, and it was no sooner out of the water than he was down again. The man immediately cried out for assistance, which was granted him as soon as possible, and they soon hauled their prize on board. His wings measured, from tip to tip, twelve feet, and his beak was seven inches in length. The captain had a dog on board, which had a fight with the bird, but the bird came out conqueror, as the bird got a good hold of the dog's tongue, and nearly pulled it out by the roots. After we had made sport enough with him, one of the men cut off his head and picked off the feathers, of which there were sufficient for a good pillow, but sea-fowl feathers are not very good, as they are always damp. I cooked the bird for dinner on Monday, and it made an excellent dinner for all hands on board.

This monstrous bird is called the Albatros, the beak of which I cleaned and kept in my chest about four years, when I was ship-wrecked, and lost all. While cutting in whales in warm climate, those large birds would flock around to assist us in taking care of the whale, although we needed none of their assistance.

At this time we were steering direct westward in 4° South Latitude, and it was very warm, although the breeze sometimes was quite refreshing to us. One day as the old *Copia* was sailing at the rate of five knots an hour, which was nearly as fast as she would go, just at supper time one of the men ran to the cabin door, and cried out at the top of his voice—

Man overboard! man overboard!"

"Can't stop for him now," replied the captain, in a very harsh manner.

The man stepped to the side of the ship again, when he saw it was the dog, then returned to the cabin door and exclaimed in a much more easy tone—

"I was mistaken; it was the dog that is overboard—*not* a man."

As soon as the captain heard that, he jumped upon deck and immediately ordered the starboard boat to be cleared away, at the same time hauling the ship to the wind. The boat was lowered away and manned in double quick time. Capt. Newell jumped into the boat, and exclaimed in a trembling voice, with tears starting from his eyes:

"Pull, my good fellows, pull."

In a few short moments they reached the dog and hauled him into the boat, being almost strangled by the salt water. When they came aboard the next thing was, to ascertain who threw the dog overboard. Not a word was uttered for a minute or two, when a Portugese boatsteerer approached the captain and informed him who it was, as this boatsteerer was a witness of the scene.—Quick as thought Capt. Newell ran to the man, and nearly killed the poor fellow.

After sailing along several degrees just south of the Equator, we took a north-west course for the Sandwich Islands.[20] We came up within a few miles of the islands at night, and in consequence of a dense fog which arose, we were obliged to heave to till morning. Morning appeared in all the splendor of a bride, but still we could not see the top of Mauna Loa. As we came near the island of Owhyhee, the ship was hove to for a pilot, and very soon we saw one coming off. As soon as he arrived the white flag was hoisted as a signal which indicated that there was no one sick on board.

The pilot was an Englishman, but had lived in the Sandwich Islands so long, and got so accustomed to their language, that we could scarcely understand him. However, he took us safe into Byron's Bay which forms the harbor of Hilo.—This bay was named by Lord Byron. Hilo is the principal town on the island of Owhyhee, the Governor of which was a native of Boston, but he was taken to the Sandwich Islands by his father when quite young. He continued to remain there after his father died. He married King Kamahamaha's daughter, when the King presented to him the whole island of Owhyhee and made him Governor of it.

After we had had our usual time of liberty on shore, we attended to our duties again, with renewed vigor and strength. The crew were fitting over the rigging, and scraping and painting ship.

On Saturday after we had had our liberty, one of the natives from the country came on board, and appeared to be very friendly, and, after a pleasant conversation, (for he understood a little English) he invited me very kindly to go home with him that night, and to go to church with him on Sunday. I accepted his invitation, and left the boy in the cabin to do my work, as he could then do it very well.

At four o'clock in the afternoon we went ashore and started on our journey. We had about ten miles to travel, and not a very good road either. We arrived at his hut at about seven o'clock, but when we got about half way, it commenced to rain, so that we were pretty thoroughly drenched before we arrived there. There was only one room in the house, and neither cellar nor attic. In the center of the room was a fire kindled upon the ground, for there was no floor, and I lay myself down upon a mat beside the fire and dried my clothes. At ten o'clock my clothes were all nearly dry, and we all retired together on a mat made of coarse grass. I slept as soundly as I ever did on the softest feather-bed.

The next morning I arose and went out to see forests of orange and bread fruit trees and sugar cane. Not long previous I was out of sight of land, but at that time I was out of sight of water. At nine o'clock we were all ready and started off for church, which was about three miles farther on. We arrived at the church, which was an old bamboo hut, without floor, seats, pulpit, or windows, before many had arrived, and I therefore took a station near the entrance, for there was no door, where I sat down upon the ground to watch the people coming in.

A most primitive manner of riding and of dress is noticed here. If a family is "well off" in this world's goods, they usually wear *some* article of clothing in vogue in the town; but most of worshipers had but little clothing, and many—both men and women—had no article of clothing, except a bit of cloth to girt about their loins. The congregation came on foot, on horse-back, on cattle, asses, or anything else which would bear them up and carry them to church; and, upon reaching the church would tie their animals to the trees which surrounded the hut.

When the congregation were called together, they "squatted" upon the ground—Indian fashion—and prepared to listen to a native preacher, who wore no shoes nor stockings and had a vest which he had probably purchased from one of the sailors, a "hickory" shirt, and a pair of coarse pantaloons made from bagging.

The services were opened by singing, by the whole congregation, among which I noticed some excellent voices, from hymn books printed in their own language. After singing, the "Mikanary," as the Kanakas called him, made what I presume was a very eloquent prayer, but could understand but little of it. Then another hymn was sung by the whole congregation; and then the sermon was preached. He spoke as I should judge very eloquently, although I could not understand much of it.

The services occupied about two hours and a half, and was listened to with close attention by the whole audience. The very stillness of death reigned during the whole time, and I have never attended church where more attention was paid to the teachings of the pastor. As I saw those ignorant people attending their rough looking churches, it reminded me of what splendid edifices we have in this country, and to be attended by so small a number. That minister, I say, was an eloquent man, for—

"Such vast impressions did his sermons make,
He always kept his flock awake."

Those who wore clothing to church, I noticed, relieved themselves of the superfluity immediately upon their arrival at home; and in no house or hut in the country would you see clothing worn.

After getting our necessary provision and water on board, the ship nicely painted, and rigging tarred so that it glistened in the sun like a new silver dollar, we had a day or two leisure time, and it then being clear, pleasant weather, we had a grand view of Mauna Loa, which is sixteen thousand feet in height, and continually covered with snow and ice; while on the splendid plains below it was scorching hot. On those plains are fine oranges, cocoanuts, pineapples, bananas and plantain, mammee-apples sugar-cane, and various other kinds of fruit.

The mammee tree generally grows fifteen or twenty feet high, with a soft stalk. The leaves are large, some of which are three feet long and one foot wide. The fruit contains a bright, yellow pulp, of a pleasant taste and fragrant smell, which is covered with a thick, tough rind.

Bread-fruit also grows in great abundance on those islands. The trees are about the size of our apple trees. It produces a fruit of a round or oval shape, and as large as a small loaf of bread, which is eaten raw or cooked.

The *Copia* sailed out from Hilo for the Arctic Ocean about the middle of March, 1853. The weather was mild and pleasant, and with a light but favorable wind, we were soon in those cold Northern Seas. We went in upon the coast of Kamseatka,[21] at Cape St. Thadeus, and there we found a liberal quantity of ice. The most of the time in that part of the world it was quite foggy, sometimes so thick that we could scarcely see from one end of the ship to the other, and accompanied by a heavy mist.

We cruised around near Cape St. Thadeus for a week or two, without any success, when we ventured a little farther north, as the summer advanced, until we came within about one hundred miles from Behring's Straits. We anchored in a snug little bay on the coast, and a boat's crew went ashore. We found bones of whales of an enormous dimension; we also found plenty of bears, deers, foxes, etc., but found but one human habitation, and *that* did not seem very much like human.

But recollect there was quite a family in that one habitation, as I should judge there were two or three families living there. The hut was built of mud, or rather a sort of clay, and partly under ground, which was covered with snow. The people appeared to be very friendly to Americans. They had some very fine furs and ivory, some of which were of the best quality, which they offered to sell very low, namely, one good fur coat for two pounds of tobacco, or a bottle of rum, (for those were their principal articles of trade, as they would not take money,) and half a pound of tobacco for a pair of walrus tusks, which would weigh six pounds, and four pounds of tobacco for a birdskin coat, which is made of the skins of all kinds of birds, with the feathers on. They are very beautiful indeed, but are covered with vermin in super-abundance.

I purchased a fur coat made of deerskin, and it was very warm; also, some ivory, and a canoe which was made of walrus-skin, and was about seven feet in length. This canoe was made perfectly air-tight, and with a place in the center of it just large enough for a man to stand up. The natives use them for catching seals, having one oar with two blades, so that a man can paddle it along at a pretty rapid rate. I paid seven pounds of tobacco for the boat, which I kept until we arrived at Hong Kong, China, where I sold it for a crape shawl which was worth fifty dollars; but the deerskin coat, as soon as we got into warm climate, was spoiled by getting wet and I threw it overboard.

We cruised along a few miles farther and anchored again,—soon after we anchored a canoe came off with some Indians[22] in it, who endeavored to make us understand that there were some white people on shore. Our captain did not apprehend their meaning, and the next day, when the old *Copia* was about to sail, having already weighed anchor and a portion of the sails were set, a boat came off again, and persuaded the captain to drop anchor again. The captain went ashore in the boat himself, and lo! and behold, there were two white men ashore,—both Americans—who had been there during the winter.

Those men were from the ship *Citizen* of New Bedford, that was wrecked a few miles north of East Cape, and had walked to this bay, a distance of one hundred miles.[23] They were natives of Camden, New Jersey. When they came aboard they resembled the Indians much more than they did our own compatriots.

I will now give you a compendious description of those men and how they came on shore in that situation: They had on sealskin caps, deerskin coats, sealskin pantaloons, sealskin moccasins, and were completely covered with vermin from head to foot. How do you think, reader, that you would like to be in that uncomfortable situation? I think you would not be pleased with it. However, the crew gave them clothing, and they were soon quite respectable.

They described their winter's life, which was certainly astonishing to the Greeks beyond all measure. Types cannot describe it. But they bore it—

"With all the tender sufferance of a friend,
As calmly as the wounded patient bears
The artist's hand that ministers his cure."

They said that their captain told them they should remain in the Arctic until "the ice was frozen to the masthead." And so he did—against his own will. Remember, Norton was his name.[24] It was on the 16th of September, when a heavy gale arose and increased rapidly, and on the morning of the 20th at three o'clock, they espied the land to the leeward, and at six o'clock they were all thrown upon the beach. Their ship went head on to the beach, when some of them jumped overboard immediately and some took it very calm. There were two persons drowned while trying to get ashore, as there was a very heavy sea running upon the beach, and one person was frozen to death after getting ashore.

In a few moments after they reached the shore the ship was broken in pieces by the sea, and in a short time it was nearly all buried in the sand. There were no inhabitants near the landing, and the crew were obliged to travel about five miles before they succeeded in finding inhabitants, and all of their clothes were wet and frozen to their bodies. However, they found the Indians very friendly, when in a day or two the storm subsided they went down to the beach, and succeeded in getting some of the provision which was partly buried in the sand, such as beef, pork, bread and molasses, and carried them to the Indian's hut. But the crew could obtain but little of the provisions, as the Indians would take it themselves and give the crew something else. They found a keg of brandy upon the beach, the head of which they broke immediately and poured the contents of it upon the beach. They said that the Indians would not even let them go to the door during the winter, through fear that they would freeze to death.

Just as soon as the weather became somewhat mild, the first officer and one of the sailors set out for the southward, and traveled until they had arrived at this bay, where we found them.

They used to go up on to a mountain, at the foot of which stood their little hut, partly under ground, and covered with snow,—there, I say, they used to go and sit and watch for the vessels that might come. And, after watching for many days they espied a ship coming in that way, but to their great astonishment, it turned its course and then sailed away. The old

Copia happened to be the first ship that came in and anchored. The first officer afterwards went on board of the ship *Bartholomew Gosnold* as first officer, and the seaman remained with us until we arrived at the Sandwich Islands, when we were informed that Capt. Norton was then captain of the ship *Northern Light*. I have neglected to state in the proper place that the *Citizen* had two thousand three hundred barrels of oil on board at the time she was wrecked.

It was anything but pleasant in that extreme cold region, for, not a tree nor a shrub could be seen. I believe that was the most gloomy place I have ever seen, without any exception. The fuel that is used in that country is the blubber of whales. The Indians go out and take a whale, tow him to the beach at flood-tide, and at ebb-tide the blubber is taken off and packed away in a hut designed for that purpose, where it remains until it is used. The Indians eat it, as I have seen an Indian with a piece weighing about two pounds in one hand and a knife in the other, cutting it off and eating it, when it smelled so bad that I could not stand within ten rods of him.—They also eat lice, for I have seen the Indians pick lice from their body and eat them. They eat none but the largest, some of which are nearly a quarter of an inch in length, for they wish to have the little ones grow.

But listen, here it was June, and not a whale had we caught since leaving the Sandwich Islands,—eight months from home and only twenty barrels of oil. However, we cruised around, and looking anxiously for whales, until the first day of July, when the man at the masthead cried out "There she blows!"

"Where away?" cried the captain, in a loud but very pleasant voice.

"Two points forward of the lee beam, sir!"

The boats were all cleared away in a twinkling and after the monster they pulled. I was shipkeeper at that time, and had a pretty tough time of it, as the whale would go first one way and then the other, so that it was hard up and hard down the wheel all the time. Finally, they succeeded in killing the whale and towed it along side of the ship. At length, the greasy substance was all on board at about three o'clock the next morning, when all except three were sent below to their berths. At seven all hands were called to get their breakfast and then get up steam.

You might, with safety, imagine that that was considerable of a sight for me, for it was the largest whale I had ever seen. This dead specimen of living monsters made two hundred and twenty-five barrels of oil, and which occupied several days in trying it out. At last, the job was finished, the oil and bone all below, and the ship cleaned.

The Fourth of July was of no account with us at that trying time, for the oil must be taken care of. We captured several smaller whales, and about the last of July we proceeded up through Behring's Straits into the Arctic Ocean. We met with but very little success in that part of the world, but we caught two whales, and that was better than most of the ships did, for there were fifty vessels in the Arctic Ocean that season, and only thirty whales were taken.

The last whale we took in the Arctic Ocean was in Latitude 69.30 north, when Capt. Newell wrote a letter, stating the size of the whale, when and where it was taken, and by what ship. The letter was put into a bottle and sealed, and then it was thrown overboard. During a part of the time in the Arctic we were surrounded by large ice-bergs, some of which were in the vicinity of an hundred feet high. We remained there beating and banging about till the 20th day of September, at which time the *Copia* was in Latitude 73° north, when we squared away for the Sandwich Islands, —just one year previous the *Citizen* was lost.

We were soon on our way back to the Sandwich Islands, there being a very favorable breeze, and our old "wash-tub" as the sailors called her, was wafted onward in the direction of the Torrid Zone. All appeared to go right, and the captain was very complacential in every respect, complimenting us often for our labor.

One dark stormy night after we passed East Cape, all hands were called to shorten sail, and as there were many ice-bergs around there, our captain deemed it prudent to carry on very little sail. It was very cold, —

> *"The night was dark and dreary,*
> *And the blast went wailing by."*

But, nevertheless, we secured the sails without much injury to them.

One day as the *Copia* was sailing along with a fine breeze, and as the captain was in good humor, he came into my pantry which I always kept looking very neat, and happened to cast his eyes upon the bill of fare which was hanging up in the pantry, when he immediately said:

"Steward, this is a good thing, this bill of fare that is hanging up here."

"Yes," said I, "haven't you seen that before? why, I have had that more than six months."

"Why, no," said he, "I have never noticed it before. It is a very good thing, however."

The bill of fare was marked in bold letters on a card about ten by eighteen inches, and read as follows:

SUNDAY,	DUFF.
MONDAY,	BEANS.
TUESDAY,	RICE.
WEDNESDAY,	POTATOES.
THURSDAY,	DUFF.
FRIDAY,	RICE.
SATURDAY,	COD-FISH.

SALT JUNK every day.

C.L. NEWHALL, Steward.

On coming down into the North Pacific, we kept a little too far to the north-east, and consequently, we were taken away out of the tradewinds, therefore we were somewhat dilatory in reaching the Sandwich Islands. But, however, the captain was not in the least puzzled in finding sufficient labor for the crew, for all the bone had to be taken up and washed and dried. It was considerable of a task to perform, as there was about fifteen thousand pounds of it. Notwithstanding the amount, it was finished.

The only death that occurred since we left home, was the death of a Kanaka, a native of the island of Molaki, one of the Sandwich Islands.

The next day after getting the bone below, we espied the top of Mauna Loa over a cloud. But we soon lost sight of it, as the cloud arose. We were then about one hundred miles from the land at the time, and there being a light wind, we did not reach Hilo until the following day, when we arrived early in the morning. And then the first thing was,

"Starboard watch get ready to go ashore!" cried Capt. Newell, in a loud but very good-natured voice, so that all could hear him. At the same time his face bore a very pleasant smile, as captains usually do after a fortunate voyage, and although we did not get as much as we anticipated, but we done better than they would average. We had then one thousand barrels on board.

I had the good fortune to belong to the Starboard watch, so I got ready as quick as possible, and at nine o'clock our watch was ashore. This time I hired a horse and traversed through the country. I first went to Pleasant Falls, which is about eight miles from Hilo, and an exceedingly bad road, but the horse was accustomed to such, for he would leap over hedges and ditches with all the ease imaginable. There was a brook that I had to cross, which ran down through a valley between two large rocks. From the top of those rocks down to the water was, I should judge, about thirty feet, and about twelve feet across from one rock to the other. As I approached it I thought certainly that I should be dashed in pieces, but, to my great joy, the horse leaped over the chasm, and appeared to do it very easy.

As I came near the falls, which was a very beautiful sight, I thought how melodious was—

"The ever-wakeful Echo here doth dwell,
The nymph of sportive mockery, that still
Hides behind every rock, in every dell,
And softly glides, unseen, from hill to hill."

After leaving Pleasant Falls I went on farther, or rather in a little different direction, towards Mt. Mauna Loa. I traveled on till I came within a few miles of the volcano, which was about forty miles from Hilo, and a few miles from the place where Capt. Cook was killed. As I was going through the country, I made some inquiries of some of the oldest inhabitants, in regard to Capt. Cook, one of which said he could recollect the time very well, but said he:

"Kanaka no eat Capt. Cook. Ouri miki."

I have often heard it stated that he was eaten by them, but that old Kanaka said they did not eat him.[25]

I found no hotels on my journey as there are in this country, and so I thought that when I was with the Kanakas I must do as the Kanakas did. As I entered their huts I seated myself upon the ground with them. They have a peculiar sort of food which they call *poi*, and is manufactured from taro, a plant which resembles the water-lily, and has large, thick, oblong roots, similar to the sweet potato, but it has a taste more like a yam. They make a hole in the ground, put stones into it, then make a large fire upon the stones, and continue it until the stones are well heated through. Then the fire is taken off and those roots or taro are placed upon the stones and covered with grass and leaves, where they remain two or three hours, when they are taken out, skinned and placed on a board, and by means of a large, heavy stick of wood they are kneaded, and moistened at the same time, so that it becomes a sort of paste similar to dough. After it is kneaded, it is put into a shell which is some like a gourd shell, and a little more water is added, where it remains until the next morning when it is fit for use.

They place this shell of poi on the ground in the center of the hut, where all sit around it and eat. One puts his fingers into it and sucks them,, and then another, and so on. They have different kinds of poi, namely, one, two, and three finger poi. One finger poi is made quite thick, so that they can eat it with one finger, and the others are thinner in proportion. I could eat it very well, if it was put into a bowl with a little sugar in it, and eat with a spoon.

I went on as far as the foot of the volcano, but did not ascend it, although many did. It was not in a very active state at the time, although I could plainly see the immense column of smoke which was continually rising from its mouth. It is fourteen thousand feet above the level of the sea. I was five days on my journey.

I had often thought of writing home, and then I thought my parents might be dead, I had delayed it so long, so that I deemed it useless for me to write, nor did I write a letter from the time I left home till I returned. I imagined at the time I was in Hilo, that I should be at home in a year, but I was not, and—

When I thought of my dear native land,
In a moment I seem'd to be there;
But alas! recollection at hand
Soon hurried me back to despair!
I might not weep, and I could not sigh,
A weight was pressing upon my breast;
A breath breathed on me witheringly,
My tears were dry, and my sighs supprest.

We sailed again from Hilo in November, 1853, first stopping at Honolulu, the principal town in the group, pleasantly situated on the island called Oahu. But we did not make a very long stop there, as we had taken in our supply of provision at Hilo. Then we sailed to the southward until we had passed Christmas and Fanning's Islands, and crossed the Equator. Both those islands are very low and sandy, having no inhabitants. And soon after crossing the Equator we kept to the westward. After sailing several degrees farther we came to one of the Ladrone Islands,[26] where the skipper purchased some pigs. There were some Americans and English on the island who had deserted vessels; some of them were naked and some were not.

Soon after leaving this island a peculiar phenomenon occurred. It was past nine in the evening, and all were asleep except the man at the wheel and myself,—some on deck and some below,—as the crew were at work pretty hard that day, and they wanted sleep, so I remained upon deck later than usual. I stood forward near the foremast at the time. All at once the firmament was lightened instantaneously, and was sufficiently light for me to observe a needle on deck. I looked forward, and saw a light descending from the sky. It was a large light, and had a tail attached, which was at least two hundred feet high. It struck in the water about ten or twelve rods from the ship, and I heard it very distinctly. No one saw it but myself, as it was forward so that the sails kept it from view of the man at the helm.

From the Ladrone Islands we steered a direct course for Hong Kong, China. We arrived in sight of land at eight o'clock, p.m., when some of the sails were furled, so that the ship would not go too fast. At three o'clock in the morning, a host of pilots (Chinese fishermen) came alongside of the ship.

"Me chin chin pilot," said one.

"Capitan, me very good," cried another; and so on.

At length, the captain consented to take one of the savage looking fellows, who resembled a monkey, only his tail was a little higher up. And we arrived at Hong Kong at seven o'clock on the 18th of January, 1854. We did not get liberty the first day, but the first night some of the boats came off, and the Chinese who came in it had some rum, or "Samshoo" as the Chinese called it. There were women, too, who came on board for the purpose of stopping over night, which is often done in China, and other countries too—I will make no exception.

The first night was a sleepless night with all hands, for they were all decidedly top-heavy, except three, including myself. Those debachees were completely mad by the poisonous, and most certainly obnoxious liquid which they drank; and

> *They were red-hot with drinking;*
> *So full of valour, that they smote the air*
> *For breathing in their faces; and beat the deck*
> *For kissing of their feet.*

And the next day they remained in their berths till nearly noon.

The first thing to be done was to wash and paint the ship, and tar the rigging, which occupied about two weeks, then we had our liberty. We arrived there just in time for the Chinese New Year, which commenced about the 20th day of January, and continued one week, during which time the gongs never ceased, and every night they displayed the town with fireworks of various description. But I was informed that in other parts of China where they had their own rulers, about two weeks' time was occupied in celebrating their New Year.

A part of the Japan fleet, under Commodore Perry, was in Hong Kong, and sailed for Japan the same day that we arrived. The names of the vessels were the Steamers *Mississippi*, *Princeton*, *Powhattan*, and the store-ship *Southampton.*—Before they sailed they gave a salute, which was followed by a salute from the battery and from the British Ships-of-War in the harbor. And as they passed out of the harbor, the American crews mounted into the rigging and gave three hearty cheers, which were followed by the British. And how melodiously those cheers echoed from ship to ship and from hill to hill.

As I have previously stated, that I did not ship for the voyage in the *Copia*, I concluded I would stop at Hong Kong and bid the *Copia* farewell. This I accomplished at last, with some difficulty between the captain and the Consul. I then went on shore to an English boarding-house kept by one James Mitchel, an Irishman, where I remained about four weeks.

I have carelessly neglected to state in a former part of this chapter, something in regard to the length of days in the Arctic Ocean, although no doubt most of my readers are aware of the fact. *In medias res*, I will give some illustration of the subject. After sailing northward three or four days, when it was very cloudy and foggy, so that the sun was invisible to us when it was above the horizon, although there was sufficient light all the time for us to see without the aid of any artificial light, it cleared away and was very pleasant.

It is customary in whale ships for the men at the masthead to come down at sunset. But this time they thought they had been up there an unusual long time, and no one ascended to relieve them, thinking it would soon be sunset. The usual time for remaining at the masthead is two hours, and after they had been up nearly three hours, the sun had descended almost to the horizon and then began to rise again, they cried out to the men on deck to come up and relieve them, because they thought the sun was not going to set again. We all thought it a pretty long day, for it was one month before it was sunset. The sun made a revolution around the horizon once in twenty-four hours, but did not go below the horizon for a month.

While writing this one pleasant evening, I saw the aurora borealis, popularly called northern lights, which reminded me of the Arctic Ocean, although they appear much more beautiful in the Arctic. This also reminds me of another strange phenomena, namely, as I saw a ship frequently at the distance of four or five miles off, an inverted ship was visible directly over it, which is caused by reflection. One afternoon as I sat on the quarter deck, I saw a ship sailing along in the sky, with the masts downward. I watched it for awhile with amazement, when I observed another ship in the water directly under it.

CHAPTER THREE

NOW, THERE I WAS, IN THAT GREAT COUNTRY, China, which, when I was a boy and went to school, I had heard older people say, was precisely under us. I believed it, and believe it still, although when I was in China the United States were under me. Thousands of miles from home, the nearest way I could go! Who would have thought it? But as I was alone in a country like that, I was not yet discouraged by the many fatigues of a mariner's life, and braved the storm, at the same time the thoughts of home kept haunting me; but with a cheerful heart I kept those thoughts silent.

During the time which I was on shore there, I traveled around the island, and visited some of the smaller villages on the opposite side of the island, which were a few miles from Hong Kong.

But before I go farther I will give a compendious description of the sights and scenes in that vast country, China. I noticed a large number of rumsellers, for nearly every house in Victoria street, the principal street in the city, was a rum-shop. I witnessed a great many hard fights with the sailors, in one of which a man was killed; and oftentimes they were badly wounded.

I traveled over hills and plains, and through beautiful valleys, where tea, oranges and lemons were growing. All was a delightful sight, but it was not in the right country for me. Several times I mounted the high cliff in the rear of the town, which is—

"So softly blending, that the cheated eye
Forgets or which is earth or which is heaven."

From the lofty cliff I could view the splendid scene before me,—the town, and the harbor well filled with ships of nearly all countries, with their colors unfurled to the breeze; and then turn to the other side and scan the broad plains and deep valleys with their beautiful blossoms and ripening fruit. Occasionally a little hut could be seen on some plantations amid the orange trees.

While I remained in Hong Kong I visited the Josh house, as the Chinese called it, which stands just in the rear of the town part way up the cliff, where the Chinese go to worship their idols. It is a fine looking building outside, built of stone and magnificently ornamented, by a serpent upon the ridge of the roof, which extends the whole length of the building; and in front of the edifice are the statues of two huge lions cut from solid rock. Various other ornaments adorn the exterior. It is somewhat rough inside, and there is no floor, but the ground is quite smooth. Around upon each side of the room are several images or idols, which are finely decorated, and in front of them lights are continually burning. In front of those dumb idols the Chinese lie down upon their face, where they remain a few moments, and then raise upon their knees, and mutter a few words which I could not understand.

"Chin chin Josh," said one as he approached me, at the same pointing at one of the idols.—He wished me to understand that they were very good.

Hong Kong is famous for cripples going about the streets, only one case of which I will mention here. As I was sitting near my boarding-house door, I witnessed a scene most awful to behold! It was an aged Chinaman who was afflicted with many sores, and like Lazarus, fed at the rich man's gate. One hand had nearly rotted from the arm, and it being warm weather the flies were devouring it as fast as possible. He had neither shoes nor hat to protect him from the scorching rays of the sun. He had a small bag suspended from his shoulder, where he kept a few coppers which he had gathered by begging. This aged man I had frequently seen at various times, and whenever I saw him I always gave him a penny to assist him. At this time his pace was slower than usual, and as he came along in front of my boarding-house, he fell upon his face in the street. He was soon raised up, but was found to be lifeless. The police then ordered some Chinese to carry the body to the police-station, which they did. The—

"Base worldlings, that despise all such as need;
Who to the needy beggar still are dumb,
Not knowing unto what themselves may come."

I went out occasionally to see the soldiers shoot at a target. I also visited the barracks and the battery, which were all well built. A Chinese ship is quite a curiosity, both in size and shape. They always have oars as well as sails attached to them. Some of them have over one hundred oars, which look very well as they all pull together. On one occasion, by an invitation, I visited an English hulk which they keep there to receive the smuggled opium that is brought in there from the northern part of China. There I saw tons upon tons of opium.

I was on board of the American Ship *Wideawake*, and while I was there a Chinese boat came alongside. One of them was a little more clumsy, or rather more unfortunate, than the others, fell overboard while clambering up the side of the ship. His companions made no effort to assist him out, and in fact, he made but very little effort himself, for he imagined the devil was after him, and he might as well go first as last. One of the *Wideawake*'s crew descended and pulled him into the boat. They were all vexed with the man for pulling him in, and even the drowning man himself was quite vexed because he had been saved. After getting him on board of their boat they immediately set fire to a large quantity of paper and threw it overboard, for the purpose of frightening the devil away, as they imagined.

After I had lived in China long enough, and had seen all I thought was worth seeing, I embarked in the whale ship *Bowditch*, Capt. Waldron, of Warren, Rhode Island. I went aboard on the 2d day of March, 1854, and sailed on the 16th. We first went to the Bonin Islands, and we were two weeks on the passage. The only accident which happened was the loss of the foretopgallant yard, which was carried away three days after we left Hong Kong.

When we arrived at the islands a boat was sent ashore for turtles, sweet potatoes, and other provision which was necessary. There I found some of the most excellent sweet potatoes I have ever seen. There are some very large turtles there, and we took six on board, one of which measured four feet in length and three feet in breadth, from which we obtained half a bushel of eggs. Our first officer came on board there, as the other one left the ship in Hong Kong.

As we sailed from the Bonin Islands we directed our course for the Japan Sea. We first sailed through China and Yellow Seas, before we reached the Sea of Japan; then we sailed along near the coast of Japan, so that we could plainly see some of the villages, which are highly noted for their grandeur. Onward we sailed—

> *O'er the glad waters of the Japan Sea,*
> *Our thoughts as boundless and our homes as free.*

The old *Bowditch* was wafted steadily along through the green waters, into the Okhotsk Sea. One evening about nine o'clock, when the moon and stars lightened the whole firmament with their bright silvery light, as we were in the northern extremity of the Japan Sea, and near the coast of China, a very warm breeze came from the land which was certainly twenty degrees warmer than that we had previously breathed. The cause of this strange wind I have never been able to ascertain, as there was no fire on shore which was visible from the ship.

A few days more of rather slow sailing found us in the Okhotsk Sea. We cruised around two or three weeks, when very suddenly, as it seemed, we discovered ourselves in a large field of ice, so that the ship was immoveable. We put out poles in order to force the ship out, one way or the other, but we labored in vain. Water was invisible to us, and as we were *invita Minerva*, Capt. Waldron ordered all sails to be furled. While thus situated we lay there quite contented, although, by a motion of the sea, the ice might have come together and swallowed us up in a moment. Our time was employed in hunting for seals upon the ice, sometimes traveling two or three miles from the ship, with a walking-stick in our hand to kill them with, as they are very easily murdered, by a light rap on the nose. The ship was fastened in that perilous situation three weeks, when, as suddenly as we were detained, we were let loose, and once more the briny wave was open before us.

After extricating ourselves from the ice, we steered for the Shanter Islands,[27] which we reached in a few days, and had tarried there but a short time when we commenced taking in the whales, as we had caught none till then since leaving China; and we were plentifully supplied with labor in cutting and trying out.

I was before the mast in the *Bowditch*, consequently my duties were a little different from the duty of a steward. I belonged to the waist-boat which was commanded by the second officer, and when working the ship this boat's crew were called the waisters, who took care of the rigging in the waist. The second officer's name was Sherman, of New London, Connecticut, and a very jolly, good-natured man he was.

We visited one island where we procured some wood and water, and on that island was a rivulet where salmon is found superabundantly. Two boats went ashore, namely, the larboard boat and waist boat, after we had got sufficient wood and water. We entered the brook, and with our hands caught upwards of three hundred in two hours, varying from two to six pounds each.

We thought, at heart, like courtly Chesterfield,
Who, after a long chase o'er hills, dales, bushes,
And what not, though he rode beyond all price,
Ask'd next day "if men ever hunted twice?"

We got about one hundred barrels of water, twenty cords of wood, and some valuable spars. After leaving that island we went to a bay on the coast of Siberia, which I will compendiously describe, as follows:

Well, in the first place, I suppose you are all well aware that it is cold weather there,—considerable colder than is experienced in the United States, and that snow is plenty all the year round. When I was there, however, the weather was comfortably mild, and there was little snow at the time. The facts I am about to relate are not very *thrilling*, still it was not very pleasant, I assure you, to live on that coast with no covering but the branches of trees, and no food except such as could be picked up about the place; and no inhabitants within about one hundred miles.

When about fifteen miles from the shore, or, as some of them said, about nine miles, the ship was anchored, and the waist boat started towards the shore, into a small bay in search of bowheads. There was no wind, and we were obliged to use our oars with all the exertions we could muster, in order to reach the bay in season to accomplish anything that day. We would pull a few moments and then stop to take breath; and at last, we arrived there, almost exhausted by the long, tedious rowing.

Soon after arriving in the bay, we made fast to a specimen of the tribe we were looking for. He snaked us around for awhile, when we killed him and went to work towing him towards the ship; but wind and tide prevailed against us, until at last, night overtook us, and a dark, dreary night it was,—the moon refused its light, and the stars were wholly invisible,—which found us there in our little boat, half a mile from the land, and no prospect of reaching the ship; and what made it still worse, was a dense fog. When we found ourselves to be so unprosperous, we anchored our boat and threw ourselves in the bottom, with the boat-sail for a blanket, where we endeavored to get a little sleep. Mr. Sherman and myself were in the aft part of the boat, and consequently were in a better position, but all of us were in a hazardous situation, for the storm might beat upon us and the waves wash over us. Soon, however, the wind commenced to blow severely, and the waves rolled high, almost taking the little boat under with every succeeding wave, and thoroughly drenching us at each fresh roll of the briny liquid.

About midnight the storm began to beat with still greater severity, and Mr. Sherman cut the rope which made fast old "bowhead," and that caused the boat to lie somewhat easier for a few speedy moments, but soon after, as the wind increased, the anchor that held the boat was broken, and we were obliged to take our oars and pull for *terra firma*. After much difficulty, and with a boat full of water, we succeeded in gaining a landing, and got our things from the boat on shore with us, although every thing was thoroughly soaked.

We immediately went to work building a hut with leaves and boughs of trees, which were chiefly spruce, pine and hemlock, with a little grass which we got in a sunny valley where the snow had melted away. There was a very heavy mist, in consequence of which the wood was all wet, and also made it very disagreeable working, and the weather being so cold, that we deemed it necessary to make a fire. How in the world we could accomplish that task of necessity was beyond our imagination. But Mr. Sherman had lived with the Indians some, and he said he could make a fire, if we could find any dry wood. That we considered a decided impossibility; but we immediately went in search for it; and after some time looking diligently, we discovered some small bits in an old hollow tree which had been blown down. At length, Mr. Sherman built a fire in the center of the hut "Indian fashion," by rubbing two sticks together.

After getting a fire started, we thought we should like something to eat, as it was then near sunset, and we had eaten nothing since the day previous. Mr. Sherman stopped in the hut to take care of the fire, and the remaining portion of us went down to the beach searching for muscles, it being ebb-tide, and we gathered about a peck. We carried them up to our little hut, and put them in the fire to roast. It was with difficulty that we left them remain a sufficient time for cooking; but we did, however. After they were nicely cooked, we eat them in the way that it is supposed the first children of men must have done, namely, with fingers and thumb. There were plenty of them, and we ate *quantum sufficit*. I assure you we were glad to eat in any manner if we could get anything to satisfy our appetite.

In this manner we lived a week, with occasionally a feast of blueberries, blackberries, or whortleberries, which grow there as well as they do in the United States. They were scarce, however, and we lived mostly upon muscles.

Two days after we went ashore, the severity of the storm was somewhat lessened, so that it was quite pleasant, and we thought we might launch the boat safely. Everything was in readiness, and the boat was moved down into the water, but just as we had got into the boat, and were about to take our oars, a tremendous swell came rolling in, which washed us all back upon the beach, and it was some time before we could recover ourselves, for we were all scattered along the beach, wet as water, and covered with sand. We got up as soon as possible, and saved the boat from being crushed to pieces, although a large hole had been made through the bottom by the experiment. So our cake was dough, for we were in just as bad a predicament as when we first landed.

A boat was sent in search of us as soon as the storm abated, and they found us at last, when we procured some tools, and mended our boat as well as practicable. We arrived on board one week from the time we went ashore. We were as happy to see those welcome faces of our old shipmates as though we had been absent for years, instead of a week. I never wish to remain in that part of the world another week without better accommodations for lodgings. The remainder of the crew had got two whales since we left, and were ready for cutting in. The one we took was driven upon the beach so far by the storm that it could not be got off. The first night as I lay in my berth meditating upon the past narrow escape, I wrote on my berth-board, with a pencil, the following stanza:

It is a week since this cellection
Has had a chance for conversation;
How glad I am beyond a question,
That I can have communication
With compatriots of our nation;
I've escaped that land of short ration,
And come on board to my old station,
Which is the place for comessation.

The *Bowditch* remained in that bay some three or four weeks, where we caught the greater portion of our bowheads, when we sailed to another bay, called the Great Shanter Bay. There, on a fine clear day, I could count a hundred whaling vessels.[28] In the morning at about four o'clock, we went off in the boats searching steadily for the black monsters, sometimes going one way and sometimes another. One morning we set off amid a thick fog, and lost our way; the fog continued during the day, and we could not find our ship. We looked, but all was vanity. We found many ships, but the *Bowditch* we had lost. As we were passing one ship, the sound of the bell told us it was nine o'clock. It was the large ship *George Washington*. Mr. Sherman hailed her, for he was acquainted with the captain, and he invited us to come aboard, which we did. They gave us some supper, and permitted us to sleep in their berths.

We arose next morning, ate some breakfast, and then started again through the fog. The fog cleared about ten o'clock, but we did not reach our ship until four in the afternoon. The other three boats did not arrive till six o'clock.

Another time we set off in a similar manner in the morning through the fog. It was foggy all day, and at night we could not find the ship, nor could we find any kind of a ship. We would pull a few moments, then stop and listen, but not a sound could we hear. The night was so dark we could not see the compass, and supposed we were pulling on a circle.

"Hark," said one, as he thought he heard a distant noise. "There is a ship off here, for I believe I can hear them pounding." At the same time pointing in the direction he heard the noise.

We listened a moment, but it was all his imagination. It was waxing late, but there was no watch among us, and we knew not what time it was. However, we were tired and sleepy enough to have some rest, so we placed ourselves on the bottom of the boat, and endeavored to sleep, but sleep was impossible in that situation, for we were shivering all night. That night seemed to me, to be sufficiently long for three or four.

Morning came at last, as clear as crystal, and where do you think we found ourselves? At sea, nearly out of sight of all the ships, and land besides. We did not arrive at the ship until about noon.

One morning at five o'clock, the bow and waist boats started off for another bay about fifty miles from the ship. It was very pleasant, and having a fine favorable wind, we set our sails, just pulling a few minutes at a time to keep ourselves warm. After we had sailed about half the distance, there was a calm, so that we were obliged to use our oars. We pulled along near the beach, and as occasion required, went ashore, looked around a few minutes, then proceeded on our voyage. As we were passing a cliff a brown bear was discovered ascending it. The third officer who commanded the bow boat, started in chase of him, but did not overtake him, as the bear could climb the best.

We arrived at our place of destination about sunset, when we went ashore, and hauled up our boats upon the bank of a river which we entered. We took the sails of both boats, and erected a tent, in the center of which a fire was kindled, where one half of us camped. One of the boats was turned over, and the remainder camped beneath it. While we were erecting our tent, an Indian hunting party came along, and camped close by us. There was also an Indian fishing party camped nearby. We rested very well during the night.

The next morning we arose, ate some bread and beef which we carried with us, and started on our return, as there were no whales in the bay. There was a fine breeze in the forenoon, which carried us about one third the distance; and then we were under the obligation of taking our oars again. We rowed along close to the beach the most of the way, so that we might go on shore *ad arbitrium*. Just after sunset we observed a ship about two miles off, and as it was black it very much resembled the *Bowditch*.

"There's the old *Bowditch* now!" cried one of the bow boat's crew.

"No; it is not her," replied Mr. Sherman.

We pulled rapidly for a few moments, to decide the question. Our boat was the fastest, and as we came sufficiently near to read the name, Mr. Sherman cried out at the top of his voice:

"There, I told you it was not the *Bowditch*. It is the *Minerva*."

We then had about six miles to go, and it was nine o'clock before we reached the ship.

Sometimes all four boats went ashore together, where we kindled a fire and had a regular muscle bake. Once we went into the mouth of a river, where there was an Indian fishing party camped. We visited their tent which was made of skins, and was very warm. The third officer had a gun with him, and he imagined himself a much better marksman than the Indians. A challenge was given and accepted, and a board was placed at a proper distance, at which they fired, but our third officer came out second best.

We remained in Great Shanter Bay until the 10th of October, when we left for the Sandwich Islands. Previous to that I assisted in composing a song to sing while heaving up anchor. The following was the chorus:

Farewell to the Shanter Islands,
Farewell! I say to thee;
Farewell to the Shanter Islands.
Likewise the Okhotsk Sea.

We were soon on our way to see the girls of Maui. There was a strong breeze right aft—

And then we saw the frothy billows fry
Under the ship, as though when she went,
That seem'd the waves were unto ivory,
Or ivory unto the waves were sent.

We passed close by Cape Lopatka, the southernmost part of Kamtschatka, being a different route from which we entered the Okhotsk Sea, —we entered through the Channel of Tartary.

I had nearly forgotten to state that we had nine Chinese on board, and we had much difficulty in getting them aloft. One day, there was a pretty strong breeze, when the man at the masthead must be relieved, and it was one of those Chinese turn to go up, but he would not go. I took a rope and told him that he *must* go.

"Hiar! no can do," he replied in a most pitiful voice. However, I was successful in driving him up.[29]

Soon after coming into the Pacific Ocean, one night about twelve o'clock, a squawl overtook us, which carried away the larboard boat *in toto*, and smashed two others that were fastened upon bearers above the quarter deck. With that one exception we arrived safe at Lahaina, Maui, one of the Sandwich Islands, with eleven hundred barrels of oil.

I was discharged at Lahaina, and went ashore the next day after we arrived. But to my great astonishment, it was not what I expected to find it. The boarding-houses were all at the lowest point of degradation. I could not remain in the cribs, and I boarded with a Kanaka in his humble hut. A steamer came in Honolulu in three days, and I then embarked for that place. Honolulu was not much better. However, I boarded there a week.

While stopping at Honolulu I dined with the King,[30] whose house was as low as the others. I also dined with other nobility of the island. One week had passed, and I shipped in the whale ship *Dover*, of New London, which was bound home. We had a fine breeze, and—

She went majestic with her swelling sails,
That gallant ship, along her watery way
Homeward she drove before the favoring gales;
Then flirting at length the streamers play,
And then they rippled with the ruffling breeze.

But alas! that ship did not take me home,—my voyage was not yet done. The ship had already sailed out of the harbor, so that I was under the necessity of going out in a boat, and two others went on board at the same time. When we arrived at the ship we found that about half of the crew had refused duty, on account of the bad condition of the ship, as she leaked excessively.

In two or three days the crew re-commenced their labors, and everything went on right for a while. But in the meantime, our captain deemed it perfectly right and proper to call at one of the Marquesas Islands, where we arrived about ten in the forenoon, and the captain with a boat's crew went ashore. The natives were barbarous and uncivilized, and they had just been at war with those upon the opposite side of the island. We went up to a group who were dancing around a fire, and gazed on them half an hour or less, when they opened the coals a little and hauled a man out from under them, which they had been cooking—a man they had killed in battle.

They soon commenced devouring him, and offered some to us, but we refused at first. They took it as an insult, and probably would have devoured us, too, had we not taken some. So to satisfy them we took a mouthful of their favorite dinner, which they had so kindly offered us. I thought as Young did, when he said,

"Their various cares in one great point combine,
The business of their lives—that is, to dine,"[31]

After eating the human flesh, and watching them a few moments, we took a slight view of the island, then went on board and sailed away.

We steered due south until we came to Latitude 40° South, when we then steered southward and eastward till we reached Cape Horn, which was about the 1st of January—mid-summer in that Latitude. Just before we came to the Cape we passed two ice-bergs, one of which was about one hundred and fifty feet in height, the other was about one hundred feet. On New Year's day the captain sent forward six bottles champagne for us.

For some days after passing the Cape, the weather was mild and salubrious, although it was rather cool. We kept around to the eastward of the Falkland Islands this time, instead of passing through the straits of La Maire. But as we came to the northward in the Latitude of Rio de la Plate, it rained almost continually, night and day, until we arrived at Rio Janeiro, which was on the 23rd day of February, 1855.

We had previously had some difficulty with the captain in regard to a large quantity of bone that had been taken before the present crew went on board. He wished to have us take the bone out of the hold and clean it. We refused to do the task, for we shipped by the month, and therefore had nothing to do with the cargo. The mate was a man that made a great ado about nothing.

As soon as we arrived at Rio Janeiro, we were all taken ashore—all that refused to take up the bone, which was fourteen—and away up to the Consul's office, where we had a thorough examination. After being examined a company of soldiers were sent for, who marched us all away (we knew not where until we arrived there) to the *calabozo*, where we remained till the ship had sailed. We had full liberty to go to any part of the building, and then, we were sufficiently strong to clear the doorway and go out; but as they always treated us with much kindness, we remained contented in the interior. We had our regular rations, which was one pound of jerked-beef and half a pint of ferino, three times a day.

There was a bathing-tub in the building, that we might bathe *ad libitum*. We also made arrangements with boarding-house keepers to bring in provision to us, so that we were bountifully supplied. Stopping in that place reminded me of the following lines:

> *"They say this is the dwelling of distress,*
> *The very mansion-house of misery;—*
> *To me, alas! it seems but just the same*
> *With that more spacious jail—the busy world."*

I have no doubt it has been a place of misery, for it had formerly been a nunnery. Sailors often came in to visit us. There was a German gentleman confined there who was convicted for counterfeiting a twenty mill rey note, on a Brazilian Bank, which was worth about eleven dollars. He was sentenced for twenty years.

When we had been in there a week or more, I thought it was plenty long enough, and suggested we should write to the Consul. I was appointed secretary, for the purpose of writing the letter. I accepted the appointment, and wrote to the Consul, as follows:

Rio Janeiro, March 1st, 1855.

Mr. Scott.—Sir:

I would ask for information, in behalf of the crew from the ship *Dover*, how long you are intending to keep us in this palace of misery; and we desire to know what we are confined for.—What have we done? If we have committed a crime sufficiently outrageous for you to convict us, we wish to know it. When we first came here we suspected the captain to be the person that imprisoned us in this contemptible mansion; but since the ship sailed, leaving us here in this place where you ought to be, we are assured that you are the very identical individual who was the cause of all this disturbance.

Please let us hear from you soon; if not, we shall proceed to higher authority. Waiting an answer, I remain your most humble servant,

C.L. NEWHALL,
Secretary for the Dover's Crew.

The letter was presented to the Consul by one of his colored servants, and March 4th the Consul's clerk came up to the *calabozo*, relieved us from that confounded old mansion of torment, and guided us away down to the Consul's office, all except two whom the captain told him were the leaders of the "mutinous gang," as he called us. Those two he kept there about a week longer; the remainder of us went to various boarding-houses. Thus ended my career in whale ships. The Consul spake in a most calumniatory manner.

Previous to our departure from the *calabozo*, we formed a petition to Congress for our relief; but, before the departure of the mail steamer we were set at liberty, and consequently it was not sent away.

CHAPTER FOUR

THERE I WAS AGAIN, YOU SEE, IN A STRANGE LAND, thousands of miles yet from old Massachusetts. I tell you, I felt as though I had not a friend in the world—I imagined every one to be my most bitter enemy. Every person I passed in the street seemed to gaze upon me with a disdainful look, and I dare not look them in the face. If ever I was in a state of discontentedness, it was at that time. But, in a few days I raised, in my own estimation, to a more lofty stand. Instead of a look of disdain, every one appeared to bear a pleasant smile when passing me. I then took much pleasure in viewing that great metropolis, and enjoyed myself by roving over the cragged hills and green valleys.

The most remarkable incident that occurred while I remained there, was proceedings on good Friday, which they highly honored. The boys had images clothed in rags, which they represented as Judas, and were dragging them through the streets during the whole day. A company of boys were hauling it along, and another company in the rear with sticks pounding it. The ships in the harbor had their yards all peaked, and a gun was fired from the man-of-war once in fifteen minutes during the day. They had an image on board to which was two ropes fastened. The ropes passed under the ship, with one end on one side and one on the other, and with a part of the crew on each end the image was hauled to and fro under the vessel. Every time it appeared above the water on each side they fired a cannon.

One day I jumped on board the ferry-boat, and crossed the Rio Grande to a town by the same name. It is not a very extensive city, but the streets are well laid out; there are some splendid buildings, and the air clear and salubrious. In that place is situated the marine hospital, which I visited. I found that they were kept very neat and clean, but had scarce enough to eat, as they said.

By an invitation from an officer—who was an Englishman—I went on board of the Brazilian frigate *Constitution* which was lying in the harbor, to observe the regulations and customs of their navy. I tarried there three days, when I was considerably wearied of my visit, for they had nothing to eat but jerked-beef and ferino. While I was on board the *Constitution* a British Admiral, a Commander, and a Captain came on board. When they departed a royal salute of twenty-one guns was fired from the *Constitution*, in honor of the British.

I often went back into the country, sometimes taking a bag with me, which I filled with oranges, cocoanuts, pine-apples, bananas, etc. Once I took a mule and rode about ten miles. I passed by some splendid, and very extensive plantations of coffee,—

"With woods of palm,
And orange groves, and fields of balm."

I oftentimes visited the Botanical garden, a sight most magnificently grand, more especially in the evening when it was lighted with lamps variegated in a very gorgeous style. In about the center is a large crystalline fountain. That park or garden stretches along the beach nearly half a mile; upon the beach is an immense platform built up, with a variegated marble floor, and surrounded with beautiful lights of different colors. I used to sit for hours upon that platform, and watch vessels coming in and going out. Hundreds of people visit that heavenly place daily.

I saw three of the Emperor's palaces, which are magnificently and sumptuously ornamented. The Emperor passed by my boarding-house door semi-weekly, in his costly coach, gold mounted, drawn by six fine horses with gold mounted harnesses. Don Pedro, the Emperor of Brazil, is a tall, well-proportioned person, with light hair and whiskers, and is about thirty years of age. One day he became vexed with his wife, as I was informed, and kicked her so that she died immediately.[32]

One evening I saw a procession of priests and others, some of which were carrying, upon tables, the full-sized figures of the twelve Apostles, and some were carrying candles, while there were a large number of little girls, with short dresses, richly trimmed with gold, and wreaths around their heads, some with banners in their hands, and some with small candles. After they had marched around the city awhile, they went into the Emperor's church, with a band of musicians to lead them. The churches are ornamented very sumptuously and beautifully.

The tradesmen in Rio Janeiro are very well divided, namely, the silk and dry goods merchants in one street, the jewelers in another, the hardware dealers in another, and so on throughout the city. In the greater portion of the city the streets are narrow and irregularly laid out, with the exception of three or four. I took delight in going out with a stranger, or a person that is easily taken astray, and take a promenade about the city. One evening I felt pretty jolly, and invited a young man out to walk with me, as I was better acquainted in that metropolis than he was. We traveled nearly all over the city, and as he was unaccustomed to traveling it was a tiresome walk for him. At about nine o'clock he thought we had better go home, and probably he would have gone, but he knew not which way to travel.

"Well," said I, "we will go soon. I want to go on through this street a little farther, then we will tack ship and go towards home."

We *were* going in the direction of our house at that time; and after walking about a quarter of a mile, and arriving at a corner, I said to him,

"Now then, I guess we'll go home," at the same time turning down the other street which was leading in the opposite direction, and consequently I was taking him away from home. We traveled in that direction about three miles, when we came to the outskirts of the city. My companion was surprised, and said,

"Hallo! here, I thought you was going home, you rascal."

"Well, I am," said I; "but I wish you to understand that I am going to give you a good view of the city first."

The moon shone very beautifully as it glided along through the azure vault of heaven, which made it a very pleasant walk for me. But as he was much fatigued and wished to retire, I turned in the direction he had long desired to go. We entered our boarding-house just as—

"The iron tongue of midnight had told twelve."

We traveled through some strange places, and saw some curious sights. Before the clock struck one we were slumbering silently in the arms of Morpheus.

One morning after having my *dejeune*, I took a tramp to the Brazilian naval hospital, passing by one of the Emperor's palaces, also his steam boat of a diminutive size, in which he enjoyed himself on pleasure excursions. The hospital is a very large structure of brick and stone. They have a commanding view of the entrance of the harbor. I passed the hospital to a fort between two lofty peaks, but having full command of the outer entrance to the harbor. I entered the fort which contained some heavy cannons and considerable of a garrison, viewed the entire fortress, then proceeded on my tour.

I then went round to a small bay, which was about eight miles from the city. I came to the final conclusion that I was far enough from home, and had better wear ship and run back, so as to return in season for tea. On my homeward tour I passed by the United States Consul's residence which is built of stone in a pleasant little grove, besides other gentlemen's mansions of splendor and magnificence.

After remaining in Rio Janeiro about six weeks I embarked on board of the Belgian bark *Lawrence*, Capt. Johnson, of Antwerp, Belgium. I sailed for Antwerp on the 2d day of April, 1855. From the time we left Rio Janeiro until we arrived at Antwerp, Sundays not excepted, there was no ceasing of labor. With a poor and scanty supply of provisions we lived through it. About three weeks previous to our arrival, on account of calm weather and contrary wind, we became short of provisions, and we were allowed only one half of our regular rations.

On the 15th day of July we arrived at Antwerp safe and sound. As soon as we arrived and the ship fastened in the dock, we immediately went ashore, for the express purpose of getting something to eat, for we had had nothing that was fit to eat since leaving Rio Janeiro. I boarded in a house kept by Frank Myres, where I remained two weeks.

The most remarkable sight I saw in Antwerp was the great Cathedral, the spire of which is four hundred and forty-one feet in height. It is an immense structure of stone, which is very rough, and probably very old. I could see it from the river Scheldt, on which Antwerp is situated, a long distance below. Belgium is a very level, and I would say the most beautiful country that I have ever traversed.

There are some very beautiful girls in Antwerp. One of them named Josephine took *my* eye to the very letter. Such beauty never before met my eyes. In fact,—

"'T were easier far to paint the hues of heaven,
When Morn, resplendent with new glory, wakes,
Or steal the varied tints by sunset given

To the gold-crested waves, the while it breaks,
Than to embody the harmonious grace
That, ever-changing, flitted o'er her face."

A good opportunity was offered, and I shipped on board of the Swedish brig *Fray*. She had been away on a voyage, and the crew had all deserted her, in consequence of which the captain was offering thirty dollars for the run to Westervik, Sweden. The brig was a very good sailing vessel, and having a favorable breeze we arrived at Westervik in eight days. We anchored at Elsinore at the entrance of the Baltic, to procure a passport. We sailed around to the westward of Gothland, and to the eastward of Olond.

On entering the harbor of Westervik, we were obliged to pass through between two large rocks or islands where there was not much room to spare on either side of the vessel. The harbor is rather small but convenient, although it is sufficiently large for all their vessels, which, at the time I was there, never exceeded twelve at once, both large and small.

Westervik is a small town, and everything appeared very quiet and modest. The principal provisions are fish, eggs, bread and milk, and a favorite dish which is called bonny-clabber. It is very good food, and especially when sweetened with sugar, it makes a *bonne bouche*. Board was one dollar a week. I remained there five weeks in a house where the whole family slept in one room, during which time I attended several social parties, and enjoyed myself very well.

One Sunday I attended church, a stone structure of an octangular form. Their religion is Lutheran. Sweden is an uneven, hilly country, and I frequently roamed over those hills to view the scenes that might present themselves. An American was scarcely ever seen in that part of the world.

As soon as the steamer *Fulton* came in from Stockholm I prepared myself to go to Copenhagen. I had taken my chest of clothes on board, and had purchased a ticket; but owing to the dilatoriness of the Governor in producing my passport, the steamer departed and I was left behind. However, it cost me nothing for my board till she arrived again in about three weeks. At that time I was ready, received my passport, and went on board. We went around near the coast, and passed between some beautiful islands covered with fine groves.

We went in to a small town, the name of which I did not learn, about nine o'clock the first evening, to take in passengers and coal. I went on shore and got some supper, as I had eaten nothing since morning. As soon as they got the coal on board we departed. The next place we came to was Carlserona, where we arrived about eleven o'clock the next day. We tarried there till two o'clock. In the meantime I went on shore, procured a dinner, and took a slight view of the village. Carlserona is chief station of the Swedish navy, and is very well fortified. The next place was Carlsham, where I went ashore and purchased some provision to take on board with me. We arrived at Copenhagen the next day about twelve o'clock.

I was detained until two o'clock when I saw the captain and ascertained where my clothes were. He informed me they were taken to the custom-house, where I found them all safe. I then went to a hotel, and remained there till the next day. While I was there I saw the statue of the late Danish and Swedish Kings. The former was on horseback and the latter was under the horse's feet. I also visited the King's palace, which is built of gray stone, two stories high. I obtained a passport, and viewed the interior of the palace. There was a Danish guide who had the kindness to conduct me into various parts of the mansionary. He conducted me to the King's apartment, in the presence of the King.[33] As I saw him he was sitting in an immense arm chair, very richly decorated with gilt and velvet of a costly style. He was a very pleasant appearing gentleman. The room was elegantly furnished with the most magnificent furniture. The city of Copenhagen is the finest, best built, and has better streets than any other city that I have seen in my travels. The houses are built mostly of stone, from two to four stories high. The people dress very plain; are plain-featured and pain-hearted. But, —

"'T is not the fairest form that holds
The mildest, purest soul within."

I went on board of a Danish steamer for Elsinore. On the passage we went close by the country palace of the King, with a splendid grove surrounding it. When we arrived at Elsinore, I went immediately to a Danish boarding-house, kept by a Dane named Ludwig, where I was sumptuously furnished with brown bread and butter three times a day, and occasionally a glass of beer, — all for five dollars a week. Mr. Ludwig was an aged man who, I should judge, was about seventy years of age. He said he had a son in New York and another in Boston, both keeping boarding-house. He expressed a strong desire for me to call upon them, but I never have been able to ascertain where they reside, consequently I have not been able to gratify his desire.

Among the most remarkable places that I visited there, was the old castle upon a point of land at the entrance of the harbor, as we came in from the Cattegat. I was informed by Mr. Ludwig that it was more than two hundred years old. I stopped at Elsinore one week, as I did not fancy the boarding very much, when I shipped aboard of the British steamship *Royal Adelaide*, of Leith, Scotland, which carried provision from Elsinore to the British and French fleet, at Russia.[34] We had one hundred and fifty head of cattle aboard, besides other provision, namely, twelve hundred bushels of sailors' hard bread, and one hundred barrels of salt beef and pork.

All was in readiness, and the ropes were loosed from the wharf. Steam was ready to bear us on our way, and Elsinore was soon left far behind. We first went to Rival where the greater portion of the fleet were lying at anchor, among which was the British Line-of-Battle ship *Duke of Wellington*. I went on board to carry a message to the Admiral, where I saw some of the Russian prisoners. I was informed by an officer that the crew exceeded fourteen hundred men. As our boat went alongside of the floating palace, the noise of the crew reminded me of an immense swarm of bees in a hollow tree. She mounted one hundred and thirty-one guns.

We then went to a bay on the coast of Finland, the name of which I do not remember, where a British frigate and two small ships of war were lying. From that bay we proceeded to Gefle in Sweden, where two frigates and several smaller ships were at anchor, besides an old frigate used for a hospital-ship. We delivered the last of our live cargo at Gefle, and then directed our course back to Elsinore, making the entire voyage in fourteen days.

I boarded with Mr. Ludwig another week, for he kept the best house in town. During my last stay in Elsinore I purchased some fire crackers, for the purpose of celebrating, but when I had nearly finished them, a policeman happened fortuitously to see me. I immediately set fire to the remainder and threw them into the street.—He came and searched me, but as he discovered none he turned and walked away. I concluded not to try the experiment again, however.

I shipped on board of the English bark *Warrington*, which was going to Pillau, Prussia, with a cargo of sugar, which was going, by way of a canal, to Russia. I found myself amongst a very bad crew, and by their extreme debauchedness, I resolved to abandon them. I spoke to the captain in relation to the matter, and told him that I disliked it very much. He immediately sent to the British Consul at Konigsburg, by whom I received my discharge.

As soon as I was discharged I went on shore, put my chest on board of a boat that was going to Konigsburg, and walked on by turnpike myself. The first night I arrived at a small town named Fishhouse, which was about twenty English miles from Pillau, being about half way from Pillau to Konigsburg. I stopped at what I should call an old fashioned tavern, with a very large bar-room. I seated myself until they had prepared supper for me, then I took a seat at the table. I had a pot of beer, some bread and butter, and sausages, which they had the impudence to call a *supper*. However, I was very much exhausted after my tedious journey, and the road being sandy made the route seem to me to be twice as long as it really was, that I could devour almost anything which was placed within my reach. After eating my supper, I concluded to retire, as it was then about nine o'clock, and I wished to rise sometime before noon the next day to resume my tour. I was lighted into a large room where there were several beds, which were neat and clean. I laid myself into one of those beds, and was very soon in the arms of Morpheus.

I arose next morning with renewed vigor and strength. At six o'clock breakfast was prepared for me, when I seated myself upon a bench at the table, on which was placed several eggs, bread and butter, dried beef, and hot coffee. I ate heartily, and then proceeded on my journey.

I traveled on through the country until I came to a tavern, where I obtained a luncheon at two o'clock in the afternoon, stopped a few moments to renew my strength, and then proceeded on in the direction of Konigsburg, where I arrived at six o'clock in the evening. I went first and took my chest from the boat which arrived before me, and then went to Hotel de Prussia, where I remained three weeks, at three dollars per week.

Some portions of Konigsburg are very beautiful and pleasant, — broad paved streets with splendid shade trees overhanging their smooth sidewalk, — while the greater portion is extremely bad, — narrow, filthy streets.

I enjoyed myself very much in attending several masquerade balls in spacious halls glistening with their beautiful chandeliers. I thought how resplendent were—

"The white arms, and the raven hair—the braids
And bracelets—swan-like bosoms—the thin robes,
Floating like light clouds 'twixt our gaze and heaven—
The many twinkling feet, so small and sylph like,
Suggesting the more secret symmetry
Of the fair forms which terminate so well."

After remaining in the city of Konigsburg a sufficient length of time, I took passage on board of the English steamship *Gertrude*, for Grimsby, England. We had an excessively rough passage, but met with no very serious accident, except the steamer came in contact with the ground when near Copenhagen, and consequently our ship was fastened there without the aid of an anchor in a situation which to me was not very pleasant, and is more easily described than realized.

It was about nine o'clock in evening when she struck, and from that time until two in the morning, all hands, passengers included, were busily engaged in endeavoring to get her free from her firm fastening. We tried experiments, namely, by carrying all the coal which was on deck from aft forward and from forward aft; by all hands going upon one side of the ship and running over to the other side; by putting on all the steam that could be got on; by dropping small anchors ahead, which was carried out in boats, and heave them in with the windlass, until at last, we were so nearly exhausted we gave it up. About seven o'clock in the morning another English steamship came along, to which we carried a cable. By the aid of the other steamship and with all the steam power we could muster, we were set free.

Upon our arrival at Grimsby, I went directly on board of a steamer which was unloading cotton, and went to work at ten shillings per day. I had only worked one day when I was informed the steam packet *Vigilant* wanted a hand, when I immediately went on shore and purchased some provision, dishes, etc., as I was to board myself. I shipped in her, and that night at eight o'clock we sailed for Hamburg. We had a very pleasant passage, and arrived at Hamburg in thirty-seven hours. As we were going up the Elbe river a large fire was seen near the river just below Hamburg upon the opposite side of the Elbe. When we arrived they informed us that it was a large bake shop which was burned.

The first operation was to scrape and oil the masts, and scrub the deck with hot water, soap and soda. The crew had nothing to do with the cargo. Every night we went ashore at Altona, where there was much sport. After our cargo was discharged and another taken in we departed for Grimsby.

On our arrival at Grimsby the same operation was to be performed, which we did with alacrity. The dock gates at Grimsby open and shut by the means of hydraulic power, and cargo is taken in and out of ships by the same power. All was in readiness and we sailed again for Hamburg. As we entered the Elbe river a dense fog came over us, so that we were obliged to go very slow, and sometimes stopped entirely. At last we arrived at our place and fastened her to piles which were driven down for that purpose. There was considerable ice in the river at the time (November) which was very inconvenient for us. But, nevertheless, we made the best of it. At that time we painted the exterior of the ship which was iron.

One afternoon I was sent on shore into Hamburg, for the purpose of purchasing a large and splendid looking-glass. I found the sought-for article after searching sometime, and started on my return. I was under the necessity of walking pretty fast in order to get out of the city before dark, for it was inclosed by a wall, through which none can pass, either in or out, without a passport. What I saw of the city was magnificent. After getting our cargo we returned again to Grimsby.

When we came to Grimsby again I deemed it prudent for me not to go to Hamburg again, as there was some danger of the vessel being frozen in. Capt. Child gratified my desire by discharging me, although he disliked to do it.

I then took the cars and went to Hull, where I tarried two days, viewing the city, the larger portion of which is very nice. They have some spacious docks which would admit a large ship. I took passage in the steamer *Water Witch*, for London, and in the course of two days more I was in that great metropolis,—that enormously large city of London,—supposed by many to be the largest in the world.

We landed near London bridge, at nine o'clock in the evening, when I went on shore, with a man whom I had become somewhat acquainted with in Hull. We traveled around the city until one o'clock, when we came to a house on Fish street hill, near where we landed. We then retired till morning. We arose about eight o'clock, took breakfast and resumed our walk through the big city.

The house where we stopped was near a large monument that was erected in memory of the great fire in London in the year 1666, having stood nearly two hundred years. I was informed that several years since a woman jumped from the top of the monument with a child in her arms and was dashed to pieces at the foot. Since that awful catastrophe occurred there has been an iron fence constructed around near the top, where people went out to view the city. But at present it is so old and the foundation so much decayed that people think it very unsafe to ascend it.

I had a distant view of St. Paul's Cathedral, which is an immense building, and by its standing upon a hill it is caused to appear, from a distance, larger than it actually is. London bridge is a very extensive bridge, lighted at night by a large number of gas lamps. Thousands of people pass over it daily.

After being wearied of the great city I walked over to Tower Hill, to a shipping office. There I observed the Tower (not the Tower of Babel, but the Tower of London.) I shipped on board of the English brigantine *Orion*, bound for ________ about six miles south of Sunderland. I first went down through Gravel Lane, through Petticoat Lane about two miles, when I came to Granville Station. While passing down Gravel Lane, I was struck with astonishment by observing two women, and what do you think they were doing? I answer silently, they were fighting upon the sidewalk. They tore off their bonnets, and nearly all their clothing was torn off from their persons. Their hair partially covered their bloody faces, and such another sight I never witnessed, either before or since that time, for—

"'T was blow to blow, disputing inch by inch,
For one would not retreat, nor t' other flinch;"

and at last a crowd gathered around and put a stop to it.

I took the cars at Granville Station and went down to Gravesend where the brigantine was lying at anchor. I arrived there at half past seven in the evening, it being dark I knew not where to find the vessel, or whithersoever to go. But, however, I stopped at an eating-house near where I supposed the vessel to be. I had been there half an hour when some men came in who were from the vessel that I was searching for. I went on board with them.

While at Gravesend we lay close by a dockyard where they were building gun-boats for the British navy. One was launched while I stopped there, and three were on the stocks nearly ready to be launched. One day we went up the river a short distance until we came near Woolwich, where they had just commenced building the big steamship *Great Eastern*. As it then appeared it did not much resemble the commencement of a ship. A little farther up I saw the Brazilian frigate *Constitution*, the ship I was three days on board of at Rio Janeiro, which had come in there for repairs. She is an American built ship.

We took on board some water, and some chalk for ballast. We sailed on the 1st day of December, and floated down the river Thames to a low, sandy beach for the purpose of taking in a little more ballast. We went in as near the beach as possible at flood tide, dropped anchor at eight o'clock in the evening and let her lay there until five o'clock, when it was ebb tide, and the vessel was on dry land. We jumped upon the beach, took out the side port, and with shovels put in about thirty tons of sand. As the tide raised so as to float the vessel, we weighed the anchor and sailed down the Thames.

As soon as we had got out of the river we had the misfortune of getting adverse winds, and soon it began to blow quite strong. We beat against the storm awhile till we gained a harbor upon the coast, and remained there till the next day, when the wind shifted in our favor. Then we tripped our anchor and proceeded on our voyage, with a cracking breeze, but it did not continue very long before we found it in the opposite direction again merely for a change. We then anchored near Yarmouth, in Yarmouth Roads. Hundred of colliers were anchored there also; and at the same time scores of loaded colliers were going towards London.

We lay in Yarmouth Roads two days, and the wind blew almost a hurricane, driving vessel after vessel upon the beach. We expected every moment that it would be our turn next. One vessel which lay near us, went ashore. Our brigantine rode the storm well with two heavy anchors, although now and then a sea broke over our bows, driving every thing before it. In two days the gale subsided somewhat, at the same time the wind veered so that it was favorable for us instead of the loaded vessels. Then we lifted our sails and anchors, when the favoring breeze sped up on our way. We met with no other difficulty on the passage, and arrived at ________ snug in the dock on the 24th day of December, the day before Christmas.

We lay idle three or four days before we could get our ballast out, but the task was very soon accomplished when they commenced upon it, for it was taken out by steam. In a small house on the wharf was a stationary engine, with a wire rope attached, and several pullies on the wharf, by this means our forty tons of ballast was taken out in two hours. After relieving ourselves of the ballast we were obliged to remain in idleness two or three days more, when they gave us a full cargo of coal.

On Christmas Eve all the inhabitants in old England were as merry as you please. We all went into the cabin with the captain, where we talked, and laughed, and sung till twelve o'clock. The merry church bells tolled nine before we left our hammocks next morning, to breathe the fragrance of the pure air which had come over us in a snow storm. At one o'clock we had a regular old English Christmas dinner,—a plum-pudding which contained more raisins and currants than flour, a leg of mutton roasted, roast beef, and a pot-pie. We ate *quantum sufficit* of the dinner which was *comme il faut*.

The storm cleared away about ten, and then the crystalline sky was as replendent as ever, and the deck was coated with a snowy sheet.

All the saloons in town took no money that day; on the contrary, they treated every person that came in. The whole of old England was in a state of merryment.

It was quite a sight to observe the vessels that were loading with coal. The coal is carried upon cars drawn by means of a stationary steam engine with a wire rope attached to the car. In this manner the coal is carried from the mines and store-houses. They will load a vessel of one hundred tons in one day; or, with the assistance of a "steam man" as they called it, a machine made of iron in the form of a man with two arms, and a large shovel attached to each arm, to trim the coal forward and aft from the hatchway, they would load a vessel of two hundred tons in one day.

New Year came at last, with sure tread, as the wheels of Time rolled on with a never-ceasing motion; and our sumptuous dinner came with it, the same bill of fare that we had for Christmas, and it seemed to me to be much better than the Christmas dinner. When—

"The big round dumpling rolled from pot,"

all were ready to take a share, and the large plum pudding which was "two raisins to one flour." Those two dinners on Christmas and New Year exceeded any that I have eaten for many a day. I will say that for old mother England.

Having received our cargo there was not work for so many men, and I being desirous of going on shore, I was discharged on the 2d day of January, 1856, although I might have continued on board had I wished to do so. I took the cars for Shields that very day; but that being a short railroad, which was only six miles, I was under obligation of leaving that train and finding another at Sunderland.

I traveled on in the direction I was told, and after going two miles I came to a tremendous great bridge. This bridge is a single span, the length of which I have forgotten, but if my recollection serves me, it is one hundred and sixty feet above the surface of the Wear river which it crosses. The arch is stone, and the space under it is sufficient for a ship to sail through. Directly after I crossed that bridge I came to the railroad station, where I remained an hour waiting for the cars. The time passed and the locomotive power was applied. We were but a few minutes going to Shields, a distance of seven miles.

As soon as I arrived in South Shields I went in search of a boarding-house, and after traveling two miles or more I succeeded in finding one which was kept by a Jew. We had four meals a day, namely, breakfast at eight o'clock, dinner at twelve, tea at five, and supper at nine, for three dollars and a half per week.

Shields is noted for its extensive coal mines, some of which extend seven miles under ground. I saw the entrance of one of them in that city, but did not enter it. Teignmouth castle in the town of Teignmouth, near Shields, is an ancient structure, and I was informed it had a spacious vault beneath it. Its walls were crumbling away, and probably, before many years have passed away, as it faces the many storms and battles, it will be numbered among the ruins of earth,—so do generation after generation crumble away like this old castle, and return to dust from whence they came.

After tarrying in Shields a few days, during which time it stormed so that I could not go out much, I shipped in the English bark *Orissa*, Capt. Ramsey, of Shields. I went on board on the 7th day of January, 1856, and sailed the 13th. The vessel was so deeply loaded that we must be very particular to go out at flood tide, because there was barely water enough on the bar to float the vessel. We were bound for the Black Sea.

The passage was very pleasant until we came down into the English

Channel, when the wind commenced to blow furiously from the south-west. We anchored in Yarmouth Roads one day, and then the wind changed to the northward, which carried us along as far as Dover. We lay at anchor near Dover two or three days, and as the wind veered so that we could steer our course or thereabout, we pulled up our "mud hook" and let her go, although it did not carry us far.

Sometimes the wind would abate for a short time, and at the same time shift around, but not to remain more than an hour or two, then shift back again. Once the wind blew and the storm beat so violently that the captain was going to return to some port with the ship. All hands were called out to wear ship, but just at that moment the wind changed to the eastward, so that it was favorable to keep on. Then the captain took courage and kept on his course; but it continued for a short time only. Finally, however, with great difficulty and inconvenience, we beat through the English Channel, into the Bay of Biscay, and there we found it still worse. The storm was fearful. Our frail bark creaked as she rose and fell with the heaving sea, and—

"Loud roared the dreadful thunder,
The rain a deluge showers;
The clouds were sent asunder,
By lightning's vivid powers.
The night both drear and dark;
Our poor devoted bark,
Till next day,
There she lay,
In the Bay of Biscay, O!"

But at length, we were favored with a fair wind, which wafted us over the Bay of Biscay, through the straits of Gibralter, and along the coast of Spain, then we had an Irishman's gale of wind, that is, right up and down, which is vulgarly called a "dead calm," for several days. During this calm weather, there was considerable hard labor, and we had a boy on board who did not like to work much. One day he went down below and laid down to sleep, but had not been there long before the mate came to the scuttle and called,

"Boy! boy! boy!"

"What, sir," said the boy, in a murmuring voice, as if half asleep.

"What are you doing there?" asked the mate very harshly.

"Catching rats!" replied the boy.

"How many have you caught?" was the next inquiry.

"When I have caught this one and two more I shall have three," was the honest reply.

The boy's ready wit saved him from punishment as it did many times.

In the Mediterranean the wind blows in various directions at the same time, and one day I recollect, we were favored with a tolerable good breeze from the west, and we espied a ship about two miles from us, going in the opposite direction with an easterly wind.

As we sailed along we came to what is called the "Drunken Sea," and I thought it was rightly named, too, for two currents came in contact with each other, which caused the sea to be in quite a rough condition. The current that runs through the straits of Gibralter, and that which runs from the Black Sea come together at that point. The water there is supposed to descend into the earth, as no sounding can be got there.

However, we passed on through the Drunken Sea as soon as practicable, through fear that it might be a catching disease and all hands become intoxicated as well as the sea. Not many days after our escape from intoxication, we arrived at the Grecian Archipelago, which, if I remember rightly, is comprised of about three hundred islands, on some of which were Grecian pirates not many years since; but at the time I passed them there were no appearance of pirates whatever.

It was some days before we could work our way through between those islands, as it was a long, crooked passage, and the wind was unfavorable for us. At length we arrived safe at the entrance of the straits of Dardanelles, and then the wind was fair, but the current was against us with a strong force; notwithstanding, the *Orissa* was driven through by a good breeze, and before long we were upon the Sea of Marmora.

On going through the Dardanelles we passed by some large forts, with tremendous big guns pointing at us. The green foliage upon the bank presented a pleasant spectacle to our view, as the trees and vines were forming their new spring dress.

As soon as we entered the Sea of Marmora we had another dead calm for a day or two, then a breeze came which carried us across the Sea of Marmora, as far as a small bay just outside of the harbor of Constantinople, where we remained about twelve hours, when a light breeze took us into the harbor. There was plenty of water, so that a pilot was not necessary. We anchored in the Bosphorus, near Scutari, at nine o'clock in the forenoon, when Capt. Ramsey went on shore in the boat. He came on board again at eleven o'clock at night, and the wind commenced to blow considerable hard from the north-east. All hands were called immediately to send down the royal and top-gallant yards.

You may easily imagine our feelings as they called us up in the dead of night, and as the wind whistled through the rigging, it cast a melancholy shudder over me. I would not have had my mother know my condition at that time for all the vessel could carry. We succeeded in getting the yards down, without any damage, about half past twelve o'clock, and went below to our hammocks again.

All hands were called at five o'clock the next morning to get the ship under way, and a tow-boat was in readiness to tow us through the Channel of Constantinople, which is commonly called the Bosphorus. We had a strong breeze and a swift current prevailing against us, consequently we did not make very much headway, however, we reached the Black Sea at ten o'clock in the forenoon.

As we entered upon the Black Sea the steamer cast off the tow-line, and we were left to depend upon our sails. There was a tolerably fair wind, when in a few days the *Orissa* was anchored at Sinope, a small town in Asiatic Turkey. Immediately after our arrival at Sinope we commenced discharging our cargo of coal into lighters, for the Turks to take on shore to the British Naval coal-yard.

We had three hundred tons of coal on board, and in the course of six days the entire mass was on shore. As this task was finished on Saturday we were allowed to go on shore Sunday to view the town, although it was not very large. We found, by viewing the outskirts of the town, many bones of men, women and children, also those of beasts, which were supposed to be the victims of the Russians. In the harbor may be seen the masts of Turkish and Russian ships of war, which had been sunk when the war commenced, for this was the first place in Turkey that the Russians invaded. The Turks have some heavy guns, one of which I saw, and was informed that it would carry a ball nine miles and do its mischief.

The week after we discharged the coal was occupied in cleaning the ship, caulking and tarring the deck. We then took in stones for ballast, and sailed for Constantinople. In a few days we arrived at Butari, at the entrance of the Bosphorus, where our ship was anchored. The captain went on shore, and remained a short time, then came on board again, when we weighed anchor and proceeded down to Constantinople. On our passage we ran the jib-boom into a house, as the houses were built upon the bank, and there being such a strong current, that it took us right in against the buildings. The building was wood and one side was laid open to the air, when the Turks ran out, astonished to see their house stove in. However, we passed without regard to what they said, and anchored near the Sultan's palace.

When we arrived at Constantinople the time for which I shipped in the *Orissa* had expired, when I left the ship with bag and baggage, and took up my residence in the city, in a boarding-house. It was extremely fortunate that they had the name of the boarding-house fastened to the building, for a stranger would as soon have called it a pig-sty as a dwelling-house. However, I managed to live there a fortnight on goat's lights, with occasionally a treat of eggs, until I was attacked with a fever.

I continued there two or three days after I was taken sick, thinking I might get better, but seeing no prospect of it, I started off one morning for the United States Consul's office. By sitting down upon the sidewalk fifteen or twenty times, to rest my aching bones, I arrived at his office at four o'clock in the afternoon, nearly exhausted by traveling so far. The Consul gave me a permit to the hospital, which was near the Consul's office, and about two miles and a half from where I boarded. He sent a servant with me to the hospital, where I lay a month with but little or no knowledge or care of what was transpiring around me.

The hospital belonged to the Prussians, and there were some good servants who took very good care of me. There were some others with the same disease, one of whom was a negro that died while I was there. For ten days after I went in there I took no nourishment whatever, and they deemed me in a very low condition. I amused myself much of the time in reading, for they had some English and American books.

As I got better and received my strength by degrees, they permitted me to go out into the back yard, to a pleasant walk beneath a splendid shade of grape-vines, where I would sit and ponder over my misfortune. I then had some more nourishing food, namely, chickens, goat meat, eggs, etc., which soon renewed my strength, and as soon as I was able I went to the Consul's office again, when I procured an order to enter a boarding-house for four days. This order I got renewed twice, and stopped at the boarding-house about two weeks, when I thought myself sufficiently recovered to go to work again.

I obtained a chance on board of the English ship *Arathusa*, that was lying near the wharf, discharging brandy, wines and lumber, that had been brought from Russia. I improved all the leisure time I had at command in seeing the sights and scenes about the metropolis, which was only on Sundays.

Occasionally I visited Pera which was about four miles from the Metropolitan Hotel where I boarded, (for I lived on shore all the time,) and during one of my visits, as I was walking along through one of the principal streets, I observed a man lying in the gutter with his throat cut from ear to ear. His features much resembled those of the Dutch, and, had probably, been brutally murdered, then left in the street for other people to take care of. Persons passed to and fro, and took no notice of it, as though it was the most common occurrance in the world. I thought of the great excitement such an affair would cause in one of our American cities. But, as I was in Turkey I thought I must do as the "Turkeys" did, and so passed on.

At another time, as I was passing over the bridge of boats, which connects Gallatta with Stamboul, or Constantinople proper, and which is about half a mile in length, I saw the dead body of a man floating along beside the bridge. There were some ugly boys upon the bridge who were *throwing stones at the corpse!* as though it was merely the dead body of a fish. Many a murder has been committed upon that bridge by those scoundrels who frequent it in the night. Turkish policemen wink at it, and, I think, usually get their share of the plunder. But this is not all. I merely mentioned those incidents because I was a witness of them. Murder is past all expiation, and, as Chapman said,

> *"Blood, though it sleeps a time, yet never dies:*
> *The gods on murd'rers fix revengeful eyes."*

It was not a very uncommon thing to hear the cry of "murder!" every night in succession, and sometimes near my boarding-house.

The bridge of boats, of which I have spoken, is constructed in this manner: One hundred boats more or less, about forty feet in length, and twelve feet in breadth, are brought alongside near each other and anchored, which extend from shore to shore. Upon those boats is built a platform with a railing at the sides, and the boats all being just the same size, it makes a strong and substantial bridge. Midway between the shores is what the Americans call a "draw." This is merely a few boats which are not anchored, and can readily be removed to allow large vessels to pass through. This is of daily occurrance.

CHAPTER FIVE

WHILE STOPPING IN CONSTANTINOPLE AMONGST the Turks, Greeks and Arabs, I nearly gave up the idea of going to sea again, and almost forgotten my old home, parents and all; in fact I abandoned a sailor's life nearly three months. I was cook in a boarding-house nearly one month, although there was but very little to cook. Goat's lights[34] and fish were the chief provision, with occasionally a few potatoes, eggs, and a little water boiled which was improperly called coffee. There were no beds to make, for there were only a few berths fastened up against the walls of the room, and every boarder must find his own bed and make it up himself, if it is done at all, besides paying the amount of about eight dollars and a half a week for board.

I disliked cooking in that way, for the boarders always came to me if there happened to be any thing that was not right, although I could help them none. There could also be seen on exhibition, all kinds of live animals, from a bed-bug up to a large sized rat, without extra charge.

In Constantinople the mosques are, of course, the same as churches in this country, only on a more gorgeous scale. In Stamboul the Mosque of St. Sophia is situated, which is built of stone of a grayish color, and is of a circular form. The main room of the mosque is in the center of the building, and all the other rooms open from it.

Surrounding the mosque is a large yard, in which are several fountains of water, and all believers wash their feet, hands and face previous to entering the mosque, for the purpose of attending to their devotions. Every mosque has one or more towers in their yard, and this one has four, to the top of which, after devotions in the mosque, every true believer mounts, and turns his face towards where Mohammed is supposed to be inclosed in a tomb, and offer up prayers to him. This is a part of their duties, and all true followers of Mohammed obey it rigorously. At all hours of the day those people may be seen wending their steps to and from those mosques; and with the remainder of the world shut out from their mind, to all appearances, they mechanically performed their accustomed devotions. I do not know the exact number of mosques in Constantinople, but should judge there were three hundred at least.

The Turks are a very strong and hardy race, as all historians mention. I have seen two Turks going along, one just ahead of the other, with poles reaching from each other's shoulders, carry a full hogshead of molasses as easily, to all appearances, as though it was a light weight, and this is a common occurrance.

The women are very small and delicate, and are never seen in the street with a male companion. They wear a white lace veil tied about their head and face when they go out for a promenade, some of whom cover their face entirely, while others cover all but their eyes and nose. When a company of two or more are walking together, you would not see two dresses of the same color. The Greek, Circassian and Georgian women are also small, and the two last-named are noted for their beauty, although I could not often have a view of their face. The Armenian women are larger, and present a good figure, but are rather smock-faced. All of the above-named inhabit Constantinople, besides many others too numerous to mention.

One day I was bent upon having a good dinner if it could be obtained in the city, for I had been there two months and had not known of such a thing as a good dinner. I went into the Hotel de Paris and called for the very best dinner they could furnish me, without any regard to expense. A servant conducted me into a splendidly furnished sitting-room, to a nice large sopia. The floor was covered by fine velvet carpet, and the large chandelier glistened as it trembled, like the many twinkling stars of the firmament. In fact, it seemed to me like a perfect paradise.

At length, a servant came and conducted me to a dining-room, to a table sumptuously spread. I found upon the table a bowl of turtle-soup, some rich cake, a bottle of champaign, a bottle of wine, besides other fixings. The entire lot only cost ten dollars. That was the only decent meal of victuals I ate in Constantinople.

Once on a time I took a horse and rode up through Pera, out of the city entirely, to the falls which was about eight miles from my boarding-house. This is a grand place of resort in summer time, and the nobility visit it with much pleasure. As I rode out of that dusty city, I thought the following sentiment of Aldrich was very appropriate in my case:

> *"Adieu, the city's ceaseless hum,*
> *The haunts of sensual life adieu!*
> *Green fields, and silent glens! we come*
> *To spend this bright spring day with you!"*

While on my tour I had the privilege of picking some figs, which were the most delicious fruit I had ever tasted.

One night I happened fortuitously to be out later than usual, and as I was passing through a narrow street, I was met by three Greeks, who detained me suddenly, and, by signs, intimated that they would relieve me of what spare change I had about me, or any other articles I might have which would prove burdensome to me. The Greek villains enforced their demands by a vigorous search of my apparel, but I was fortunate enough to leave my change at home that night, and they were rewarded by finding empty pockets. It was all in my good luck that it remained safe in the house. But, however, they were determined to have *something*, and so took my hat, coat and boots. I all the while feeling exceeding thankful that they were kind enough to leave my head, which was of no importance except to the owner, and they probably thought the same.

I thought upon the wild character of the place, and upon the pleasure of paying eight dollars and a half per week for miserable board, when I concluded to ship in an English iron ship, which was bound for Taganrog, Russia. On the 26th day of August, 1856, we hove up anchor, and was soon under way for our place of destination. We first touched at Theodosia, Russia, where, by the rules of the country, we were obliged to lay in quarantine eight days. We then touched at Kertch, a small town north of Theodosia, which was the first town the British and French troops captured in the Eastern war. After leaving Kertch we passed through the straits of Enikale, into the Sea of Azof, and then sailed the entire length of that sea, to Taganrog, at the mouth of the river Don.

On our arrival at Taganrog we ascertained that our cake was dough on getting a cargo, owing to the war which caused an extreme scarcity of the article (wheat,) and we were under the painful necessity of coming away without a cargo.

At Taganrog was a massacre by the troops, and I saw several churches and other buildings with large holes in their walls, which were made by cannon balls. When the war first commenced many citizens abandoned the town, for the interior of the country, and at the time I was there some of them were returning to resume their accustomed labors, as usual.

Returning from Taganrog we passed close by the harbor of Balaklava, where the masts of the sunken vessels might be seen projecting from the water, some of which the Russians sunk to prevent the British from entering the harbor. At Sevastopol, I understood, they sunk the largest ship in their navy, for the purpose of stopping the entrance of the harbor.

When we arrived at Butari on our return, we remained four days, and meanwhile the captain went down to Constantinople, to receive orders from England by telegraph. As soon as he returned the ship was taken to Constantinople to stop a few days.

Our ship was anchored close by the Sultan's new palace, which is a magnificent structure of marble, richly ornamented with porphyry and stucco work. The gates at the entrance of the yard are highly ornamented with rich bronze, so that they glistened in the sun, and almost dazzled one's eyes to look at them. I received information from good authority that there was one gate in the interior of the palace which was solid gold, and I should judge by the outward appearance of the palace that it might be so, although we cannot always tell what there is in the interior by outward appearances.

The palace fronts on the Bosphorus, and has a fine marble pavement in front. Upon this pavement is about twenty lamps, the posts of which are carved marble statues of women.

The Sultan or Grand Seignor of Turkey was a young man, who was at that time, I was informed, about twenty-seven years of age.[36] He is small in person, and is very sumptuously dressed. The boat in which he crosses the Bosphorus, to the seraglio and the harem, is a very long boat, richly gilded from stem to stern.

The palace withal is a splendid piece of human architecture, and while—

"The princely dome, the column and the arch,
The sculptur'd marble, and the breathing gold,"

met the gazer's eye, the remainder of the city looked as though it were poverty-stricken, although when at about a mile from the city, it appeared like a beautiful city, and—

"So the blue summit of some mountain height,
Wrapt in gay clouds, deludes the distant sight;
But as with gazing eyes we draw more near,
Fades the false scene and rough rocks appear."

As we passed from the scene of architecture, we drifted down with the current into the Sea of Marmora, and soon entered upon the beautiful prospect along the banks of the Dardanelles, that was heavy-laden with grapes and figs,

"Where olives glow with sunny dyes,
Like maidens' cheeks when blushes rise;
Where with huge figs the branches bend,
Where clusters from the vine distend."

From this we passed along down to the Grecian Achipelago, and it being autumn those isles presented a beautiful scenery. And as we sailed along between them, I thought how beautiful and pleasant they were.

When we arrived at Malta they informed us that four days previous an earthquake had visited them, and they said, as we could see, that the island had sunk about two feet. We were also informed that a town in Cyprus was entirely destroyed together with about two thousand of its inhabitants. At the time of the earthquake, we were just coming around Cape Matapan from the Archipelago, but did not realize the shock.

As our ship was to be hove down and repaired, which would probably occupy two or three months, the captain did not want the crew, so we were all discharged except two. We immediately went on shore to a boarding-house, where I remained three weeks before I could procure a berth on board of another ship. Meanwhile, I traveled over the greater part of the island, which is solid sand-stone. I saw many ancient curiosities, one of which was St. John's Church, a very ancient building. I visited it and viewed, with wonder and admiration, its beautiful walls and galleries; its gilded altars and statues; and its spacious vaults, which contained the bodies of many eminent men, some of whom thrived long centuries before us.

This old church contains a gate of solid silver, and once contained a gate of solid gold, but when Bonaparte captured the island he carried it away. Immediately after the ship sailed a storm came, and the ship was lost, together with their prize. In the place of the golden gate is a gilded one, which is very beautiful. There is a town clock upon the building which gives the hour, day of the week, and day of the month,—a valuable clock. There are hundreds of visitors gathered daily to view that ancient church.

Malta is also noted for being the place where St. Paul was shipwrecked, but I was not able to ascertain the exact place where he went ashore. The island abounds with prickley-pears, an exceedingly delicious fruit; pomegranites also grow there, but not plenty. There is no soil upon the island except what has been carried there. It is the principal stopping place for steamers when going to the Black Sea. Valletta, the chief town on the island, is said to be the strongest fortified place in the world, and they are making it still stronger.

At last I embarked on board the American ship *Sarah Boyd*, Capt. Percy, of Bath, and on the 26th day of November, 1856, we sailed from Malta for Girgenti, Sicily. It was on Sunday morning, and while—

> *"Day glimmered in the east, and the white moon*
> *Hung like a vapor in the cloudless sky,"*

there was a light breeze from the north-east, but as night approached, the breeze freshened and shifted more to the northward, until the next morning, it blew tremendously which continued four days. The wind was dead ahead, consequently we could gain no *ground*. One night, I think it was the second after we left Malta, the wind changed very suddenly to the southward, and the ship was taken all aback, almost taking her down stern foremost. However, we got her around to the wind again, but had no sooner done so than the wind shifted again to the northward, as if we were doomed for a watery grave. We got her to the wind the second time, without going to the bottom.

After the gale there was a dead calm, and this with head wind, when there was any wind, made our passage last nine days, going a distance of *sixty miles*.

When we arrived at Gergenti we anchored two miles from shore, when we threw all our ballast overboard and went in half a mile nearer. Then we commenced taking in brimstone, and after receiving about one hundred tons the breeze again freshened. The wind soon blew so furiously that the boats could not come out with brimstone.

The wind was south-west, which would take us towards the shore.

While we were endeavoring to ride out the storm in safety, with one anchor down, the cable broke which held it, and before we had time to let go the other anchor we were driven nearly a mile toward the rocky shore.

The other anchor did not hold the ship then, as the wind raged tremendously, and the chain was all out, which was ninety fathoms, still it dragged. Then the part of a chain that was left from the lost anchor was shackled on and payed out, in all, making one hundred and ten fathoms. It was then three o'clock in the afternoon, and the anchor seemed to hold fast.

The gale did not appear to increase any more, but the sea rose rapidly, and braking over our bows, drenching us to the skin at each fresh roll of the waves. Night came on apace, and it was dark—intensely dark.

> *"O, treach'rous night!*
> *Thou lend'st thy ready veil to every treason,*
> *And teeming mischiefs thrive beneath thy shade."*

I had an excellent opportunity to ponder over these lines, for it was my watch from midnight till one o'clock, and it seemed as though at night the evil spirit stalked about and concocted mischief. We were then driven so that we were no more than half a mile from shore, and at half past two the fearful cry rang through the ship, that the other cable had broken, leaving us all adrift, with the prospect of being driven upon the rocks of Sicily in a few moments. I had gone below to my hammock, but could not sleep. When I heard the alarm I immediately arose, and just as I arrived upon deck I heard the order of,

"Get sail on her as quick as possible!" cried the mate in stentorian tones.

The sails were set with great difficulty, and a few moments elapsed in silence, except the loud whistling of the wind through the rigging. Soon the more melancholy cry was heard,

"There she strikes! there she strikes!" and at the same time the captain yelled in a voice of thunder,

"Cut away the mizzenmast!"

The shrouds were cut and the mast broke about fifteen feet above the deck.

"Cut away the fore and mainmasts!" was the next order on the programme.

The rigging was all cut, but the masts were so tough they would not break, until they were cut nearly one-third through, and then those huge sticks fell with a crash over the side of the ship. We expected every moment that the vessel would be dashed to pieces upon the rocks, but fortunately, she went in between them, which none save the All-seeing eye could have seen, where, at four o'clock, she lay a little easier. She was then about half full of water.

It then being light enough to see, with partial distinctness, objects upon the shore, and soon a boat was seen coming from the beach, to which a rope was thrown, and to our great satisfaction and thankfulness, we succeeded, at six o'clock, in reaching *terra firma*, totally exhausted with our efforts to save our lives. Pen cannot describe the scenes of that night (December 18th,) but as one of the party, the recollections of the dangers of the night will never be effaced from my memory.

The United States Consul immediately found a comfortable place for us to live, where we remained four weeks. In the meantime I traveled around some. I went back in the country twenty miles, so that I could see the smoke from Mt. Etna distinctly, and on the same tour I visited a sulphur mine, which is very beautiful in the interior. I also visited the ruins of the ancient city of Aggrigentum, which was said to be destroyed 304 years before Christ, and still the ancient walls were there, but were gradually crumbling to dust. There were olive trees growing up among them, so as to partially conceal them from the distant eye.

At the end of four weeks we all went on board of the bark *Henry Shelton*, Capt. Burr, which was lying in the dock taking a cargo of brimstone and sumac. After receiving a part of a cargo, we sailed to Palermo, to get the remainder of the cargo, which consisted of oranges, lemons, walnuts, almonds, and sumac, and on the 15th of January, 1857, we sailed from Girgenti. After two days sailing we arrived at Palermo, and when the cargo had been stowed away which occupied about two weeks' time, we were allowed a day or two on shore, which time we occupied in viewing the town. Palermo is a fine city, well-built, as the buildings are chiefly stone, and most of the streets are paved well.

Everything being in readiness, we sailed for the United States on the 13th day of February. We had a fine breeze, and went along pleasantly until we had come about half way, when the wind turned against us, and the remainder of the passage was very rough, so that we were fifty-four days on the passage. We came in sight of Long Island on a cold frosty morning, but nevertheless, it was a lovely sight, and—

Once more on the deck I stand,
And cast a wishful eye
Upon my own native land,
Where I have longed to lie.

We arrived in New York on the 8th of April, but not wishing to go home at that time I stopped two weeks, then shipped on board of the schooner *John Boston*, for Savannah, Georgia. We had a very pleasant passage, with the exception of a storm off Cape Hatteras. However, we arrived there without much injury, discharged the cargo, and took in a cargo of cotton. Having taken a cargo, we returned to New York, and then I went on shore, where I remained two weeks,—while they were unloading and loading the vessel—then I went on board again to go back to Savannah. We had about the same kind of a passage as before, and when we arrived at that time it was very warm weather, so that we could not work at the cargo; and slaves from shore were hired, so we had nothing to do. The cargo being discharged and another taken in, we sailed again for New York, where we arrived on the 9th day of July, 1857, when I resolved never to leave the United States again, although I have almost got started several times, and my friends have persuaded me not to go; and—

"Oh! who can tell, save he whose heart hath tried,
And danc'd in triumph o'er the waters wide,
The exulting sense—the pulse's maddening play,
That thrill the wanderer of the trackless way."

THE END.

BIBLIOGRAPHY

Anderson, Charles Robert. *Melville in the South Seas.* 1939. Reprint, rev. ed. New York: Dover Publications Inc., 1966.

Barnacle, Captain [pseud. tor C.M. Newell]. *Leaves from an Old Log: Pehe Nu-e, the Tiger Whale of the Pacific.* Boston: D. Lothrop and Company, 1877.

Davis, William M. *Nimrod of the Sea: or The American Whaleman.* New York: Harper & Brothers, Publishers, 1874.

Ely, Ben-Ezra Stiles. *"There She Blows": A Narative of a Whaling Voyage, in the Indian and South Atlantic Oceans.* 1849. Reprint. The American Maritime Library, Vol. III. Ed. by Curtis Dahl. Middletown, Conn.: Published for the Marine Historical Association, Incorporated, by Wesleyan University Press, 1971.

Fagg, John E. *Latin America: A General History.* New York: The Macmillan Company; and London: Collier-Macmillan, Ltd., 1963.

Federal Writers' Project of the Works Progress Administration. *Whaling Masters.* American Guide Series. New Bedford: Old Dartmouth Historical Society, 1938.

Hawes, Charles Boardman. *Whaling.* Garden City, New York. Doubleday, Page & Company, 1926.

Hegarty, Reginald B., comp. *Addendum to "Starbuck" and "Whaling Masters," New Bedford Customs District.* New Bedford: New Bedford Free Public Library, 1964.

Hohman, Elmo Paul. *The American Whaleman: A Study of Life and Labor in the Whaling Industry.* New York, London and Toronto: Longmans, Green and Co., 1928.

Martin, Kenneth R. "American Whaleships in the Mariana Islands." *Guam Recorder*, 9, (1979), 3-9.

Miller, Pamela A. *And the Whale Is Ours: Creative Writing of American Whalemen.* Boston: David R. Godine, Publisher; and Sharon, Mass.: The Kendall Whaling Museum, 1979.

National Archives Project. *Ship Registers of New Bedford, Massachusetts.* Compiled by the Survey of Federal Archives, Division of Professional and Service Projects, Work Projects Administration. 3 vols. Boston: The National Archives Project, 1940.

Newell, C.M. *The Isle of Palms: Adventures while Wrecking for Gold, Encounter with a Mad Whale, Battle with a Devile Fish, and Capture of a Mermaid.* Boston: De Wolfe, Fiske & co., 1888.

Newhall, Charles L. *The Record of My Ancestry.* Southbridge, Mass.: Herald Power Print, 1899.

Scammon, Charles M. *The Marine Mammals of the North-west Coast of North America, Described and Illustrated: Together with an Account of the American Whale-Fishery.* 1874. Reprint. New York: Dover Publications, Inc., 1968.

Starbuck, Alexander. *History of the American Whale Fishery from its Earliest Inception to the Year 1876.* 1878. Reprint. 2 vols. New York: Argosy-Antiquarian Ltd., 1964.

Syme, Ronald. *The Travels of Captain Cook.* New York and Toronto: McGraw-Hill Book Co., 1971.

Torrey, William. *Torrey's Narrative: or, The Life and Adventures of William Torrey* etc. Boston: Press of A.J. Wright, 1848.

Waldron, Ambrose. Journal kept on board ship *Bowditch* of Warren, 1852-1854. The Nicholson Whaling Collection, Providence Public Library, Providence, Rhode island.

Wood, Dennis, Abstracts of Whaling Voyages From the United States, 1831-1873. 7 vols. Melville Whaling Room , New Bedford Free Public Library, New Bedford, Massachusetts.

Notes

1. Southbridge, Ma.; Printed by the Auther [sic], 1859.
2. Elmo Paul Hohman, *The American Whaleman: A Study of Life and Labor in the Whaling Industry* (New York, London and Toronto: Longmans, Green and Co., 1928), pp. 58-62.
3. Charles Roberts Anderson, *Melville in the South Seas* (1939; reprint, rev. ed., New York: Dover Publications, Inc., 1966), p. 12.
4. Pamela A. Miller, *And the Whale Is Ours: Creative Writing of American Whalemen* (Boston: David R. Godine, Publisher: and Sharon, Ma.: The Kendall Whaling Museum, 1979), p. 16.
5. Captain Henry F. Gifford (Reginald B. Hegarty, comp., *Addendum to "Starbuck" and "Whaling Masters," New Bedford Customs District* [New Bedford: New Bedford Free Public Library, 1964], p. 63).
6. The *Leonidas* was the thirty-three years old, having been built at Fairhaven in 1816 (National Archives Project, *Ship Registers of New Bedford, Massachusetts* [compiled by the Survey of Federal Archives, Division of Professional and Service Projects, Work Projects Administration, 3 vols., Boston: The National Archives Project, 1940], I, 183, nos. 1806 and 1818).
7. The Azores.
8. Trindade, a Brazilian island ten degrees east of the mainland.
9. It should be added that Tristan da Cunha is located in the mid-South Atlantic.
10. The right whale's "roaring" was caused by the violent expulsion of its breath. Such "bellowing" has been described by other whalemen (Charles M. Scammon, *The Marine Mammals of the North-western Coast of North America, Described and Illustrated* etc. [1874; reprint, New York: Dover Publications, Inc., 1968], p.69; Ben-Ezra Stiles Ely, *"There She Blows": A Narrative of a Whaling Voyage, in the Indian and South Atlantic Oceans* [1849; reprint, The American Maritime Library, Vol. III, ed. by Curtis Dahl, Middletown, Ct.: Published for the Marine Historical Association, Incorporated, by Wesleyan University Press, 1971], p. 53),
11. Baleen.
12. Saint-Paul, an uninhabited Indian Ocean island consisting of a partially submerged volcanic crater, was frequently resorted to because of its abundant fish stocks.
13. Reunion, one of the Mascarene Islands.
14. Anjouan, one of the Comoro Islands.
15. Faustin I.
16. Possibly the later author of curious, maritime adventure novels for youngsters, three of which were inspired by Melville's *Moby-Dick: Leaves from an Old Log' Pehe Nu-e, the Tiger Whale of the Pacific* (Boston: D. Lothrop and Company, 1877); *The Voyage of the Fleetwing: A Narrative of Love, Wreck and Whaling Adventures* (Boston: DeWolfe, Fiske and Company, 1886); and *The Isle of Palms: Adventures while Wrecking for Gold, Encounter with a Mad Whale* etc. (ibid., 1888).
17. Sao Jorge, one of the Azores.

18. Diego Ramirez.
19. Juan Fernandez is also known as Mas a Tierra. Selkirk's residence on the island was fictionalized by Daniel Defoe in *Robinson Crusoe*
20. The Hawaiian Islands.
21. Kamchatka.
22. Eskimos.
23. The *Citizen was lost 300 miles north of East Cape, Siberia, 14 October 1853, with six deaths (Alexander Starbuck, History of the American Whale Fishery from its Earliest Inception to the year 1876* [1878; reprint, 2 vols., New York: Argosy-Antiquarian Ltd., 1964], II, 478-479).
24. Thomas A. Norton, a veteran New Bedford skipper (Federal Writers' Project of the Works Progress Administration, *Whaling Masters* [American Guide Series, New Bedford: Old Dartmouth Historical Society, 1938], p. 211).
25. Cook's remains were cremated, not eaten (Ronald Syme, *The Travels of Captain Cook* (New York and Toronto: McGraw Hill Book Co., 1971), p. 175.
26. The Marianas. The *Copia* may have stopped at Anathan Guguan or Pagan, islands known to have been the resort of beachcombers (Kenneth R. Martin, "American Whaleships in the Mariana Islands," *Guam Recorder,* 9 (1979) p. 6.
27. The Shantar Islands are off the Siberian Coast in the western Sea of Okhotsk.
28. This surprising figure is confirmed in Ambrose, Waldon, Journal kept on board ship *Bowditch* of Warren, 1852-1854 (The Nicholson Whaling Collection, Providence Public Library, Providence, Rhode Island). 29 Juy 1854.
29. In later life, Newhall described his nine Chinese shipmates as "very good sailors generally" (Charles L. Newhall, *The Record of My Ancestry* (Southbridge, Herald Power Plant, 1899), p. 27.
30. Kamehameha III.
31. It requires a leap of literary faith to imagine Newhall's poetic musings as he munches long pig; but for less esthetical account of a similar situation, see William Torrey, *Torrey's Narrative: or, The Life and Adventures of William Torrey* etc. Boston: Press A.J. Wright, 1848), pp. 128-129. Had Newhall read Torrey?
32. Pedro II was happily married. The story Newhall heard involved the unpopular Pedro I, who, many Brazilians believed, had committed such an outrage upon his wife, (John Edwin Fagg, *Latin America A General History* [New York: The Macmillan Company; and London: Collier-Macmillan, Ltd., 1963], p. 528).
33. Frederick VII.
34. At this time the Crimean War was reaching its climax.
35. Lungs.
36. Abdul Mejid I.
37. Newhall, *The Record of My Ancestry,* p. 36.
38. *Ibid.*, pp. 37-41.
39. *Ibid.*, p. 38.
40. *Ibid.*, pp. 22-23, 45.
41. *Ibid.*, pp. 43-45.
42. *Ibid.*, p. 23.

INDEX

D

E

F

G

H

I

J

K

L

M

COLOPHON

This edition of Charles Newhall's LIFE ON THE WAVE *was printed in the workshop of Glen Adams, which is located in the sleepy country village of Fairfield, southern Spokane County, Washington State. The type was set by Miss Bobi Pearson using a Compugraphic 48 computer photosetter. The setting is in 12 on 14 Baskerville on 30 pica lines. The running heads and page numbers are in 18 point Diamond Inlaid, a Victorian face rarely seen today. The paper stock is Hammermill Opaque in a vellum finish. General book design and letterpress printing in color is by Glen Adams. Camera-darkroom work is by Evelyn Foote Clausen, who also folded the sheets using a Pitney Bowes multi-sheet folder. The text pages were printed offset by Robert LaTendresse who also did the stripping and opaque work. The sheets were printed on a 777CD Hamada offset press. The book was indexed by Father Edward J. Kowrach. Considerable assistance on the project was given by Kenneth B. Martin of Walpole, Maine, who wrote the introduction and supplied footnotes. My thanks are extended to Mr. Martin for his helpfulness and courtesy. Some correction work was done on the Newhall book by Miss Kristy Frisbie. The binding is by William Bosch of Oakesdale. All typographical errors that sneaked past us are included at no extra charge. This was a fun project. We had no special trouble with the work.*

5/92 Jg

UNITED STATES F

The U.S.S. *Minnesota*. Newhall served on board this ve
Hampton Roads. Courtesy of The Kendall Whaling M